Introduction to Developmental Biology

Introduction to Developmental Biology

Francis Collins

R CALLISTO REFERENCE

www.callistoreference.com

Callisto Reference,
118-35 Queens Blvd., Suite 400,
Forest Hills, NY 11375, USA

Visit us on the World Wide Web at:
www.callistoreference.com

ISBN: 978-1-64116-559-4 (Hardback)

Cataloging-in-Publication Data

Introduction to developmental biology / Francis Collins.
    p. cm.
Includes bibliographical references and index.
ISBN 978-1-64116-559-4
1. Developmental biology. 2. Biology. 3. Growth. I. Collins, Francis.
QH491 .I58 2022
571.8--dc23

TABLE OF CONTENTS

The purpose of this book is to help students understand the fundamental concepts of this discipline. It is designed to motivate students to learn and prosper. I am grateful for the support of my colleagues. I would also like to acknowledge the encouragement of my family.

The study of the processes through which plants and animals grow and develop is referred to as developmental biology. It encompasses various areas of study such as biology of regeneration, metamorphosis, asexual reproduction as well as the growth of stem cells in the adult organisms. The developmental processes of organisms are divided into two major categories, namely, cell differentiation and regeneration. The process in which different functional cell types arise during development is known as cell differentiation. The ability to regrow a missing part is known as regeneration. Some of the other processes studied within this field are regional specification, morphogenesis and growth. This book unfolds the innovative aspects of developmental biology which will be crucial for the progress of this field in the future. The topics included herein on this subject are of utmost significance and bound to provide incredible insights to readers. Coherent flow of topics, student-friendly language and extensive use of examples make this book an invaluable source of knowledge.

A foreword for all the chapters is provided below:

Chapter – Introduction

The study of the process which includes the development and growth of plants and animals is referred to as developmental biology. It encompasses various areas of study such as the biology of asexual reproduction, regeneration, metamorphosis as well as the growth and differentiation of stem cells in adult organisms. This is an introductory chapter which will introduce briefly all the significant aspects of developmental biology.

Chapter – Basic Concepts of Developmental Biology

Some of the most important concepts of developmental biology are embryology, aging, embryogenesis, metamerism, metamorphosis, ontogeny, teratology and regeneration. This chapter has been carefully written to provide an easy understanding of these concepts of developmental biology.

Chapter – Development Processes in Animals

The process which leads to the formation of a new animal and which begins with the derivation of cells from one or more parent individuals is known as animal development. It involves the processes such as mating, fertilization, blastulation, implantation, etc. The topics elaborated in this chapter will help in gaining a better perspective about these processes related to animal development.

Chapter – Development Processes in Plants

Plant development refers to the production of new tissues and structures by plants throughout their life from meristems. Some of the most important processes of plant development are plant reproduction, photomorphogenesis, morphological variation, etc. This chapter discusses in detail these processes related to plant development.

Chapter – Evolutionary Biology

The subfield of biology which deals with the study of the evolutionary processes that produce the diversity of life on Earth is referred to as evolutionary biology. It includes processes such as natural selection, common descent and speciation. This chapter closely examines these processes of evolution as well as adaptation, heredity, genotype and genetic variation.

Francis Collins

TABLE OF CONTENTS

Introduction

• Developmental Biology

The study of the process which includes the development and growth of plants and animals is referred to as developmental biology. It encompasses various areas of study such as the biology of asexual reproduction, regeneration, metamorphosis as well as the growth and differentiation of stem cells in adult organisms. This is an introductory chapter which will introduce briefly all the significant aspects of developmental biology.

Developmental Biology is the study of progressive changes in size, shape, and function during the life of an organism by which its genetic potentials (genotype) are translated into functioning mature systems (phenotype). Most modern philosophical outlooks would consider that development of some kind or other characterizes all things, in both the physical and biological worlds. Such points of view go back to the very earliest days of philosophy.

Among the pre-Socratic philosophers of Greek Ionia, half a millennium before Christ, some, like Heracleitus, believed that all natural things are constantly changing. In contrast, others, of whom Democritus is perhaps the prime example, suggested that the world is made up by the changing combinations of atoms, which themselves remain unaltered, not subject to change or development. The early period of post-Renaissance European science may be regarded as dominated by this latter atomistic view, which reached its fullest development in the period between Newton's laws of physics and Dalton's atomic theory of chemistry in the early 19th century. This outlook was never easily reconciled with the observations of biologists, and in the last hundred years a series of discoveries in the physical sciences have combined to swing opinion back toward the Heracleitan emphasis on the importance of process and development. The atom, which seemed so unalterable to Dalton, has proved to be divisible after all, and to maintain its identity only by processes of interaction between a number of component subatomic particles, which themselves must in certain aspects be regarded as processes rather than matter. Albert Einstein's theory of relativity showed that time and space are united in continuum, which implies that all things are involved in time; that is to say, in development.

The philosophers who charted the transition from the nondevelopmental view, for which time was an accidental and inessential element, were Henri Bergson and, in particular, Alfred North Whitehead. Karl Marx and Friedrich Engels, with their insistence on the difference between dialectical and mechanical materialism, may be regarded as other important innovators of this trend, although the generality of their philosophy was somewhat compromised by the political context in which it was placed and the rigidity with which their later followers have interpreted it.

Philosophies of the Heracleitan type, which emphasize process and development, provide much more appropriate frameworks for biology than do philosophies of the atomistic kind. Living organisms confront biologists with changes of various kinds, all of which could be regarded as in some sense developmental; however, biologists have found it convenient to distinguish the changes and to use the word development for only one of them. Biological development can be defined as the series of progressive, nonrepetitive changes that occur during the life history of an organism. The kernel of this definition is to contrast development with, on the one hand, the essentially repetitive chemical changes involved in the maintenance of the body, which constitute "metabolism," and on the other hand, with the longer term changes, which, while nonrepetitive, involve the sequence of several or many life histories, and which constitute evolution.

As with most formal definitions, these distinctions cannot always be applied strictly to the real world. In the viruses, for instance, and even in bacteria, it is difficult to make a distinction between metabolism and development, since the metabolic activity of a virus particle consists of little more than the development of new virus particles. In certain other cases, the distinction between development and evolution becomes blurred: the concept of an individual organism with a definite life history may be very difficult to apply in plants that reproduce by vegetative division, the breaking off of a part that can grow into another complete plant. The possibilities for debate that arise in these special cases, however, do not in any way invalidate the general usefulness of the distinctions as conventionally made in biology.

The Scope of Development

All organisms, including the very simplest, consist of two components, distinguished by a German biologist, August Weismann, at the end of the 19th century, as the "germ plasm" and the "soma." The germ plasm consists of the essential elements, or genes, passed on from one generation to the next, and the soma consists of the body that may be produced as the organism develops. In more modern terms, Weismann's germ plasm is identified with DNA (deoxyribonucleic acid), which carries, encoded in the complex structure of its molecule, the instructions necessary for the synthesis of the other compounds of the organism and their assembly into the appropriate structures. It is this whole collection of other compounds (proteins, fats, carbohydrates, and others) and their arrangement as a metabolically functioning organism that constitutes the soma. Biological development encompasses, therefore, all the processes concerned with implementing the instructions contained in the DNA. Those instructions can only be carried out by an appropriate executive machinery, the first phase of which is provided by the cell that carries the DNA into the next generation: in animals and plants by the fertilized egg cell; in viruses by the cell infected. In life histories that have more than a minimal degree of complexity, the executive machinery itself becomes modified as the genetic instructions are gradually put into operation, and new mechanisms of protein synthesis are brought into functional condition. The fundamental problem of developmental biology is to understand the interplay between the genetic instructions and the mechanisms by which those instructions are carried out.

In the language of genetics the word genotype is used to indicate the hereditary instructions passed on from one generation to another in the genes, while phenotype is the term given to the functioning organisms produced by those instructions. Biological development, therefore, consists of the production of phenotypes. The point made in the last paragraph is that the formation of the phenotype of one generation depends on the functioning of part of the phenotype of the previous generation (*e.g.*, egg cell), as the mechanism that begins the interpretation of the instructions contained in the new organism's genotype.

Types of Development

In the entire realm of organisms, many different modes of development are found, the most important categories of which can be discussed as pairs of contrasting types.

Quantitative and Qualitative Development

Development may amount to no more than a quantitative change (usually an increase) in a system that remains essentially unaltered. Qualitative development involves an alteration in the nature of the system. Pure examples of the first type are difficult to find. Approximations to it occur when an animal or plant has attained a structure with the full complement of organs; it then appears to increase only in size, that is to say, quantitatively. This would be a period of simple growth. A closer examination nearly always shows that the system is also undergoing some qualitative change, however. A human infant at birth, for example, already has its full complement of organs, but the ensuing developmental period up to adulthood involves not only growth but also processes of maturation that involve qualitative as well as quantitative changes. Perhaps the most uncomplicated examples of quantitative development occur in certain simple plants and animals. Flatworms, for example, may become reduced in size when starved but increase in size again when provided with suitable nutrition; they thus undergo quantitative changes. Even in these cases, however, it is found that the constituent organs do not always merely become reduced in size but may actually suffer the loss of certain parts.

Progressive and Regressive Development

The normal processes of development in the majority of plants and animals may be considered progressive since they lead to increases in size and complexity and to the addition of new elements to the system. As already indicated, some organisms, when placed in adverse conditions, may undergo regressive changes, both in size and complexity. Such regressive changes are a part of the normal life history of certain organisms. Characteristically, these are species in which the organism at an early stage develops a relatively complex structure that enables it to be motile, and later adopts a form of life for which motility is no longer a necessity. A good example is that of the barnacles, a group of marine crustaceans in which the egg at first develops into a motile larva that soon settles down and becomes firmly attached to a solid underwater surface. The barnacle then loses many of the organs characteristic of the motile phase and develops into its familiar stationary form.

There are a number of other examples, particularly in groups in which the adults adopt a parasitic form of life, especially within the digestive system or other tissues of a host animal, from which they have only to absorb their nutriment without having to move or to possess suitable organs for capturing prey. In such cases the early developmental period is characterized by progression toward more complex forms followed by a period of regression in which many of these organs may be lost. During this regressive period certain components of the organism (*i.e.,* those concerned with functioning as a sessile or parasitic form) may undergo progressive development at the same time as the other organs are regressing.

Single-phase and Multi-phase Development

The most familiar organisms, including man, undergo a single-phase development; the organs that appear at early stages persist throughout the whole of life. There are many kinds of animals that develop one or more larval stages adapted to a life different from that of the adult. Perhaps the best known of these is the common frog. The egg first develops into a tadpole, which is provided with a large muscular tail by which it swims. The tadpole eventually undergoes a change of form, or metamorphosis. This involves the regression and resorption of the tail and the growth of the limbs. During this time the rest of the body of the tadpole undergoes less profound changes; the organs persist but undergo relatively far-reaching progressive changes. In other animals, the alteration between the larval and the adult forms may be much more drastic. The egg of a sea urchin, for instance, at first develops to a small larva (the pluteus), which is completely unlike that of the adult. During metamorphosis nearly all the structures of the pluteus disappear; the five-rayed adult develops from a very small rudiment within the larva. In other groups of marine invertebrates, there may be successive larval stages before the adult form appears.

Plants in general appear to exhibit a type of development related in a general way to the multi-phased development in animals, although rather different from it in essence. This is called the "alternation of generations." The majority of higher plants possess two sets of similar chromosomes in each of their cells, that is to say they are diploid ($2n$), as are most higher animals. But in sexual reproduction, diploid cells undergo a reduction division so as to form precursors of the sex cells, which are haploid—*i.e.,* they contain only one set of chromosomes. In animals these cells develop directly into the sex cells—egg and sperm—which unite in fertilization. In plants the haploid cells undergo some developmental processes before the functioning sex cells are produced. The products of this development are spoken of as the "haploid generation." In most higher plants the haploid development is quite reduced, so that the haploid individuals contain only a few nuclei—those associated with the pollen tube on the male side and a few associated with the egg on the female side. In some lower plants, however, such as mosses and ferns, the haploid development may be much more extensive and give rise to quite sizable separate plants. In such cases a species contains two kinds of individuals, produced by different types of developmental processes controlled, however, by the same genotype. This may be compared with the multiphasing development of larval forms in animals. The situation in plants, however, is characterized by the two forms of the organism having different chromosomal constitutions—haploid and diploid—whereas the larval forms and the adult of an animal species have the same chromosomal constitution.

Structural and Functional Development

These two categories cannot be regarded as a pair of opposites they are two aspects of all processes of biological development and can be separated only conceptually, and for purposes of convenience of description. Function is the capacity of the biological system to carry out operations. At the level of the organism, these operations include walking, swimming, eating, digesting, etc.; at the cell level, typical functions are respiring, contracting, conducting nervous impulses, secreting hormones, etc.; and at the molecular level, all functions depend on the production of enzymes, coded by particular genes. Structure encompasses all parts of the organism capable of carrying out functions localized within the body of the organism and arranged in some particular spatial pattern. Contractile cells, for example, are grouped together to form muscle, and other cells are grouped together to form elements of the skeleton; both the muscles and the skeletal elements have definite spatial relations to each other.

These two aspects of development—function and structure—are not opposed to each other in any way. On the contrary, it is obvious that the higher level functions are clearly dependent on the proper structural relations and functions of cell systems. Even at the basic cellular or molecular levels, secretion or nervous conduction essentially depends upon the proper structural relation of the subcellular elements. It is, however, often convenient to focus discussion on one or other of these two aspects of development; for instance, a study may be made into the developmental processes that bring about the production of hemoglobin or insulin by a certain kind of cell, without at the moment being concerned with structural problems. Or again, the focus may be on the results of a certain process by which a mass of cells develops into a typical hand with five digits. In such an inquiry the structural aspects are paramount.

Normal and Abnormal Development

If a number of fertilized eggs of a given species are provided with conditions that enable them to develop at all, they will, with extraordinary regularity, develop into exceedingly similar adult organisms. The range of conditions they can tolerate is rather wide, and the similarity of the end products surprisingly complete. There are, indeed, good grounds for recognizing what must be considered normal development. The situation is perhaps more marked in animals than in plants, since the plants produced from a given batch of seed under a variety of environmental conditions often present considerably greater variation than is commonly found among animals. Even among plants, however, the differences produced by different conditions of cultivation are usually no more than quantitative differences in size and number of such organs as leaves and flowers, so that an individual can be described as well or poorly developed rather than as normally or abnormally developed. It is only in relatively few cases that a plant develops in quite different ways under two different conditions, neither of which can be considered abnormal or normal. In certain aquatic plants, for instance, the shape of the emergent leaves is different from the leaves that develop underwater. In such cases the plant actually has two normal forms of development.

It is possible, of course, to produce abnormal organisms by submitting a developing system to stimuli not usually encountered in a normal environment, such as certain chemicals. The presence of unusual genes also may result in deviations from the normal processes of development. In the vast majority of cases such abnormalities can be regarded as resulting from failure to carry out fully the normal processes of development. Functional abnormality in the adult consists in the failure of the system to produce a certain enzyme or functional cell type; a structural abnormality consists in the unusual appearance of certain component elements or in their arrangement in incompletely realized patterns. It is extremely rare to find examples in which the abnormality consists in the addition of a new enzyme not produced in normal development, or the formation of a new structural pattern of the elements.

One very important type of development that, from some points of view, can be considered as an exception to the rule that abnormal development is nearly always retrogressive, is carcinogenesis, the production of tumours. Carcinogenesis involves a change in the developmental behaviour of a group of cells. Initially, it often involves a loss of some of the functional and structural characteristics that previously appeared in the cells. It is commonly followed, however, by the assumption of new properties, which however untoward they may be for the host animal, must be considered as a progressive type: the cells often grow faster and multiply sooner than the noncancerous cells, for example. Furthermore, the cells may undergo a sequence of changes in character and in their arrangement within the tumour. All these features can be regarded in a developmental sense as progressive.

In view of the great rarity of cases of abnormal development that lead to progressive changes, it seems to follow that the organs produced during the normal development of any given species actually exhaust all the potentialities of its genotype for the production of orderly functional structures. It appears that the only abnormal developments that can be produced are either displacements of normal organs, or inadequacies in carrying out normal processes, or the initiation of progressive but quite disorderly processes, as in the production of tumours.

General Systems of Development

Development of Single-celled Organisms

In viruses, activities consist in the production, aided by the machinery of a host cell, of units for building new virus or phage particles: development is simply the assemblage of these constituent units.

In the next higher grade of biological organization, the organism consists of a single cell. Many single-celled algae produce special forms of cells that correspond to the sex cells, or gametes; these cells may unite in fertilization, the resulting fertilized egg, or zygote, undergoing a short period of development. In many other single-celled organisms, however, reproduction takes place by the simple division of an original cell into two daughter cells. In such forms, development normally is part of the process of subdivision. It involves the remodelling of the parent cell into two smaller cells, which are then separated by the division. Something similar must, of course, be involved in the division of cells of higher organisms also. In many single-celled organisms, however, the cell contains a number of defined parts, which are arranged in very definite ways, so that the process of remodelling is very striking and easily observed. This is so, for instance, with ciliated protozoans, in which the cortex is provided with a large number of hairlike cilia or other appendages, arranged in precise patterns, and often with such other structures as a mouth or a gullet. These structures are reproduced in two identical but smaller copies during cell division. This does not necessarily imply that no other developmental processes are possible. The process of regeneration of parts removed occurs quite independently of cell division, for example.

Open and Closed Systems of Development

There is a marked difference between the general system of development in multicellular plants and multicellular animals. In a plant, certain groups of cells retain throughout the whole life of the plant an embryonic capability to give rise to many types of cells. These regions, known as meristems, occur at the growing tips of branches and roots and as a cylindrical sheath around the stem. They consist of rapidly dividing cells capable of assembling into groups that form buds from which may arise new stems, leaves, flowers, or roots.

By contrast, most animals have no special regions that retain an embryonic character. In most forms, the whole egg, and the whole collection of cells immediately derived from it, take part in the developmental processes and form parts of the developing embryo. In some forms that go through a number of larval stages, the development of certain cells is interrupted at an early stage, and they are set aside and resume their development to form a later type of larva, or to form the adult after the larval stages are completed. An example would be the imaginal buds of some insects. The cells of these buds cannot be regarded as retaining a fully embryonic character comparable to that of the plant meristems, since they cannot perform all the developmental processes but only those involved in the production of the particular late-larval or adult structure for which they have been set aside. In general, then, plants remain embryonic in character, capable as it were of starting again from the beginning to carry out the entire developmental process. Their development is, in this sense, "open." Most animals, on the other hand, lack persistently embryonic cells of this kind, and their development may be characterized as "closed." (There may be certain exceptions to this in very simple forms, such as flatworms, in which certain cells called neoblasts seem able to participate in any type of development; these cells are usually scattered throughout the body, and the major developmental processes that bring into being the general form of the organism cannot be attributed to them, as the development of the plant can be attributed to the meristems.)

Blastogenesis versus Embryogenesis

Some animals possess a second system of development, in contrast to the "closed" embryonic system. In its most fully developed form, this system consists in remodelling a portion of the parental body into a new organism without any involvement of eggs or sperm. In an adult hydra, a microscopic aquatic animal, a portion of the body may begin to grow exceptionally fast; its cells differentiate into the various cell types and become molded into the constituent organs to build up a new individual identical to the parent. The group of cells responsible for this behaviour is, in its early stages, referred to as a bud, or blastema. Before they become activated these cells may appear quite indistinguishable from the other cells of the body and betray no embryonic capability comparable to the meristems of plants.

In some higher organisms, including certain insects, reptiles, and amphibians, incomplete but still fairly extensive new developments of a similar kind may take place. They require the stimulus of an injury, however, which may involve the removal of part of the normal body. The usual result is a new development to regenerate, or replace, the missing part. The first stage in such regenerative processes consists in the formation of a blastema, that is, a group of rapidly dividing cells that shows little sign of cellular specialization. The evidence indicates that they may not arise, as was once thought, from persisting embryonic cells scattered within the adult body, but instead are formed of cells near the position of the injury. These cells lose their normal adult character and become capable of developing into most of the tissues required to replace the parts removed by the injury.

Development from a blastema, or blastogenesis, presents many contrasts to embryogenesis, the normal form of development from a fertilized egg. In blastogenesis, tissues that, during embryonic development, appear in sequence one after another, may be formed simultaneously and without any obvious sequential relations. Very little, however, is as yet understood about the mechanisms by which the various tissues within the blastema become differentiated from one another. It may well be that these mechanisms are more similar to those found in embryonic development than appears at first sight.

Constituent Processes of Development

Growth

Developing systems normally increase in size, at least during part of their development. "Growth" is a general term used to cover this phenomenon. It comprises two main aspects: (1) increase in cell numbers by cell division and (2) increase in cell size. These two processes may in some examples occur quite separately from each other; for instance, cells in certain rapidly growing tissues (*e.g.*, the connective tissue or blood-forming systems in vertebrates) may increase greatly in number, while the cells remain approximately the same size. Alternatively, in some organs (*e.g.*, the salivary glands of insects) the cells may increase greatly while remaining the same in number, each cell becoming enlarged, or hypertrophied. In such greatly enlarged cells there is often duplication of the genes, involving an increase in the DNA content of the nucleus, although no cell division takes place, and the nucleus continues as a single body, although with a multiplied, or "polyploid," set of chromosomes.

In very many cases, however, the growth of an organ depends on increases both in cell number and in cell size. The relative importance of these two processes has yet to be properly investigated. One case that has been well studied is the size of the wings of the fruit fly *Drosophila*. The number of cells in the wing can be easily determined, since each bears a single hair that can be seen and counted in simple microscopic preparations. It has been found that there is an accommodation of factors: if there is an unusually large number of cells, these may be somewhat smaller than usual, so that the total size of the wing remains relatively unchanged.

Perhaps the major theoretical difficulty in the concept of growth is that it is a quantitative notion attached to an ill-defined entity. Growth is an increase in size; but size of what? If a cell or organ increases in volume merely by the absorption of water, or by the laying down of a mineral substance such as calcium carbonate, is this to be regarded as growth or not?

Morphogenesis

Morphogenesis refers to all those processes by which parts of a developing system come to have a definite shape or to occupy particular relative positions in space. It may be regarded as the architecture of development. Morphogenetic processes involve the movement of parts of the developing system from one place to another in space, and therefore involve the action of physical forces, in contrast to processes of differentiation, which require only chemical operations. Although in practice the physical and chemical processes of development normally proceed in close connection.

There is an enormous variety of different kinds of structures within living organisms. They occur at all levels of size, from an elephant's trunk to organelles within a cell, visible only with the electron microscope. There is still no satisfactory classification of the great range of processes by which these structures are brought into being.

Morphogenesis by Differential Growth

After their initiation, the various organs and regions of an organism may increase in size at different rates. Such processes of differential growth will change the overall shape of the body in which they occur. Processes of this kind take place very commonly in animals, particularly in the later stages of development. They are of major importance in the morphogenesis of plants, where the overall shape of the plant, the shape of individual leaves, and so on, depends primarily on the rates of growth of such component elements as the stems, the lateral shoots, and the vein and intervein material in leaves. In both animals and plants, such growth processes are greatly influenced by a variety of hormones. It is probable that factors internal to individual cells also always play a role.

Although differential growth may produce striking alterations in the general shape of organisms, these effects should probably be considered as somewhat superficial, since they only modify a basic pattern laid down by other processes. In a plant, for instance, the fundamental pattern is determined by the arrangement of the lateral buds around the central growing stem; whether these buds then grow fast or slowly relative to the stem is a secondary matter, however striking its results may be.

Morphogenetic Fields

Many fundamental processes of pattern formation (*e.g.*, the arrangement of lateral buds in growing plants) occur within areas or three-dimensional masses of tissue that show no obvious indications of where the various elements in the pattern will arise until they actually appear. Such masses of tissue, in which a pattern appears, have been spoken of as "fields." This word was originally used in the early years of the 20th century by German authors who suggested an analogy between biological morphogenetic fields and such physical entities as magnetic or electromagnetic fields. The biological field is a description, but not an explanation, of the way in which the developing system behaves. The system develops as though each cell or subunit within it possessed "positional information" that specifies its location within the field and a set of instructions that lays down the developmental behaviour appropriate to each position.

There have been several attempts to account for the nature of the positional information and of the corresponding instructions. The oldest and best known of these is the gradient hypothesis. In many fields there is some region that is in some way "dominant," so that the field appears as though organized around it. It is suggested that this region has a high concentration of some substance or activity, which falls off in a graded way throughout the rest of the field. The main deficiency of the hypothesis is that no one has yet succeeded in identifying satisfactorily the variables distributed in the gradients. Attempts to suppose that they are gradients of metabolic activity have, on investigation, always run into difficulties that can only be solved by defining metabolic activity in terms that reduce the hypothesis to a circular one in which metabolic activity is defined as that which is distributed in the gradient.

Recently, a new suggestion has been advanced concerning position information. Most processes within cells normally involve negative feedback control systems. These systems have a tendency to oscillate, or fluctuate regularly. In fact, any aspect of cell metabolism may be basically oscillatory in character; the cycle of cell growth and division may be only one example of a much more widespread phenomenon. The substances involved in these oscillations are likely to include diffusible molecules capable of influencing the behaviour of nearby cells. It is easy to envisage the possibility that there might be localized regions with oscillations of higher frequency or greater amplitude that act as centres from which trains of waves are radiated in all directions. It has been suggested that positional information is specified in terms of differences in phase between two or more such trains of transmitted oscillations.

Certain types of field phenomenon may involve an amplification of stochastic (random) variations. In systems containing a number of substances, with certain suitable rates of reaction and diffusion, chance variation on either side of an initial condition of equilibrium may become amplified both in amplitude and in the area involved. In this way, the processes may give rise to a pattern of differentiated areas, distributed in arrangements that depend on the boundary conditions.

Morphogenesis by the Self-assembly of Units

Complex structures may arise from the interaction between units that have characteristics such that they can fit together in a certain way. This is particularly appropriate for morphogenesis at the simple level of molecules or cells. Units such as the atoms of carbon, hydrogen, oxygen, nitrogen, and so on, can assemble themselves into orderly molecular structures, and larger molecules, such as those of tropocollagen, or protein subunits in general, can assemble themselves into complexes whose structure is dependent on localized and directional intermolecular forces. It seems that such comparatively large entities as the units that come together to form the head structures of bacteriophages or bacterial flagella are capable of orderly self-assembly, but the chemical forces that give rise to the interunit bonds are still little understood.

Processes that fall into the same general category as self-assembly may occur within aggregates of cells. The units that self-assemble are the cells themselves. Interaction and aggregation may be allowed to occur in assemblages of cells of one or more different kinds. In such cases it is commonly found that the originally isolated cells tend to adhere to one another, at first more or less at random and independently of their character, but later they become rearranged into a number of regions consisting of cells of a single kind. When the cells in the initial collection differ in two different characteristics, for instance in species and organ of origin, the assortment in some cases brings together cells from the same organ, in other cases cells from the same species. Mixtures of chick and mouse cells, for instance, reassort themselves into groups derived from the same organ, whereas cells from two different species of amphibia sort out into groups from the same species more or less independently of organ type.

This morphogenetic process probably has only a restricted application to the formation of structures in normal development, in which only in a few tissues (*e.g.*, the connective system) do cells ever pass through a free stage in which they are not in intimate contact with other cells, and cells of different origin do not normally become intermingled so as to call for processes of reassortment. To explain normal morphogenetic processes of plants and animals one must look to the results that can be produced by the differential behaviour of cells that remain in constant close contact with one another. Several authors have shown how striking morphogenetic changes could be produced within a mass of cells that remain in contact, but that undergo changes in the intensity of adhesion between neighbouring cells, in the area of surface in the proportion to cell volume, and so on.

Differentiation

Differentiation is simply the process of becoming different. If, in connection with biological development, morphogenesis is set aside as a component for separate consideration, there are two distinct types of differentiation. In the first type, a part of a developing system will change in character as time passes; for instance, a part of the mesoderm, starting as embryonic cells with little internal features, gradually develops striated myofilaments, and with a lapse of time develops into a fully formed muscle fibre. In the second type, space rather than time is involved; for instance, other cells within the same mass of embryonic mesoderm may start to lay down an external matrix around them and eventually develop into cartilage. In development, differentiation in time involves the production of the characteristic features of the adult tissues, and is referred to as histogenesis. Differentiation in space involves an initially similar (homogeneous) mass of tissue becoming separated into different regions and is referred to as regionalization.

Histogenesis involves the synthesis of a number of new protein species according to an appropriate timetable. The most easily characterized are those proteins formed in a relatively late stage of histogenesis, such as myosin and actin in muscle cells. The synthesis of proteins is under the control of genes, and the problem of histogenesis essentially reduces to that of the genetic mechanisms that direct protein synthesis.

Regionalization is concerned with the appearance of differences between various parts of what is at first a homogeneous, or nearly homogeneous, mass. It is a prelude to histogenesis, which then proceeds in various directions in the different regions so demarcated Unlike morphogenesis, regionalization need not involve any change in the overall spatial shape of the tissues undergoing it. Regionalization falls rather into the type of process for which field theories have been invoked.

Control and Integration of Development

Phenomenological Aspects

One of the most striking characteristics of all developmental systems is a tendency to produce a normal end result in spite of injuries or abnormalities that may have affected the system in earlier stages. In many cases, perhaps in most, only injuries inflicted during a certain restricted period of development can be fully compensated for. During such periods the system is said to be capable of regulation or the restoration of normality.

Developmental regulation is often discussed in terms of homeostasis, or regulatory mechanisms. Many systems, including biological ones, exhibit a tendency to return to initial equilibrium once they are diverted from it. A developing system is, by definition, always changing in time, moving along some defined time trajectory, from an initial stage, such as a fertilized egg, through various larval stages to adulthood, and finally to senescence. The regulation that occurs in such systems is a regulation not back to an initial stable equilibrium, as in homeostasis, but to some future stretch of the time trajectory. The appropriate word to describe this process is homeorhesis, which means the restoration of a flow.

A second major phenomenological characteristic of development is that the end state attained is not unitary but can be analyzed into a number of different organs and tissues. The overall time trajectory of this system can, therefore, also be analyzed into a number of component trajectories, each leading to one or another of the end products that can be distinguished in the later stages. A major discovery of the early experiments on developing systems was that, in many cases at least, the different time trajectories diverge from one another relatively suddenly during some short period of development, which usually occurs well before any visible signs of divergence can be seen microscopically or by any other available means of analysis. The most dramatic and influential example of this was provided by studies on the development of the amphibian egg at the time of gastrulation, or formation of a hollow ball of cells. At this time the lower hemisphere of the embryo will be pushed inward (invaginated) to develop into the mesoderm and endoderm, and the upper hemisphere will remain on the surface, expanding in area to cover the whole embryo. Approximately one-third of the upper hemisphere will develop into the nervous system and the remainder into the skin. During the period when these morphogenetic movements of invagination and expansion are occurring, a process takes place by which a portion of the upper hemisphere enters a trajectory toward neural tissue and another part enters a trajectory leading to epidermal development. This process of determination of developmental pathways happens relatively quickly, during a period when the cells of the two different regions appear superficially alike. The occurrence of the determination can in fact be demonstrated only experimentally. Before it occurs, any part of the hemisphere can develop either into neural tissure or into skin. After it has happened, each part can develop only into one or the other of these alternatives.

It is clear that an adequate theory of development has to account not only for the processes by which a developing system moves along its appropriate time trajectory, but also for the nature of the processes by which the trajectories diverge from one another and become fixed or determined in the developing cells.

The determined state can be transmitted through many cell generations. An example of this transmission can be seen in Drosophila flies. The imaginal buds of Drosophila are small packets of cells that become separated from the main body of the embryo in the early stages of development. They persist throughout larval life and then enter into the differentiation of adult characteristics when stimulated to do so by the hormones secreted at the time of pupation. These pupation hormones disappear from the body of the adult insect, and imaginal buds transplanted into the body cavity of an adult undergo many cell generations, but they do not show any signs of differentiating into the specific tissues of the corresponding adult organ. After many generations of proliferation, however, the cells can be transplanted back into a larva ready to pupate; they thus submit to the pupation hormones and differentiation occurs. Through many generations of proliferation the cells have retained the determination as to which adult organ they will develop into when the pupation hormones become available.

Attempts to identify the determining agent have not yet been successful. Experiments on amphibian eggs, however, have given rise to one important general conclusion; namely, that the process of determination can take place only during a certain period of development, in which the cells of the upper half of the amphibian egg are poised between the two alternatives of development into neural tissue or into skin. They are said at this time to be "competent" for one or the other of these types of development. While they are in this state, and only while they are in it, a variety of external agents can switch them into one or the other of the possible pathways. Such a situation may be contrasted with one in which the cells were neutral, or featureless, and required then an external agent to transmit to them the quality of becoming nervous tissue or of becoming skin. This would mean that the reacting cells required information or instructions to be added to them from outside. Such a situation is not characteristic of biological development. Both in highly developed organisms such as amphibians and in simpler ones such as bacteria, the external agents act only as a releaser that switches on one or another process for which all of the necessary information is already incorporated in the cells concerned.

Analytical Aspects

The existence of these developmental phenomena was realized in the first third of this century. During this period, biologists had no clear notion of the fundamental concepts needed to explain development. Developmental biologists, or embryologists, attempted to account for their observations by means of ill-defined notions, such as "potencies" or "organ-forming substances," or by referring to cellular properties that are real enough but obviously in themselves complex and essentially secondary in nature, such as cellular adhesiveness, the capacity of cell surfaces to differentially absorb certain substances, and so on. It was only gradually that developmental biologists came to realize the importance of the demonstration by genetics that nearly all the instructions required for the building of a new organism are contained in the genes that come together during fertilization, and that the small additional amount of information, contained primarily in the ovum, is itself a product of genetic instructions provided in the body of the mother in which the ovum is produced. The fundamental problems of the theory of development are, therefore, to understand how these units interact with one another to form more complex mechanisms that bring about the cellular or tissue behaviours of the different types of developing systems.

In the development of the neural system of vertebrates, for example, a great many genes must be active in controlling the synthesis of particular proteins. In the formation of the wing of a Drosophila, the activity of some 20 or 30 genes has been definitely demonstrated, and certainly many more are involved. The action of all these genes, however, must be considered to form a network involving many types of feedback and other interactive loops, the overall result of which is a product in which many components are present in precisely defined concentrations; and further, the developmental process leading to this end result must be buffered or stabilized, in the sense that if the process is diverted from its normal course at an early stage, it returns to some later stage of the normal trajectory. The realization that the basic units of development are genes indicates that a stabilized time trajectory involves the action of tens, if not hundreds, of genes. The realization that biological development is fundamentally an expression of the controlled activities of genes has finally resolved one of the old philosophical controversies about the nature of development, between preformation and epigenesis. The former supposed that, at the initiation of development, for instance in the fertilized egg, the system already contained some representative of every organ that would eventually put in an appearance. The vindicated theory of epigenesis, on the other hand, supposed that later appearing entities were produced during the course of development.

The modern interpretation of epigenesis is that the initial stage of development does contain certain entities with well-defined properties, namely the genes. These do not, however, represent directly the later formed organs, which arise by the gradual interaction and progressive unfolding of the properties of groups of genes.

One of the major problems confronting modern developmental biology—namely, the nature of "determination"—requires an understanding of how genes are "primed" to enter into activity when an appropriate stimulus is given. The state of priming presumably has to apply to quite a large number of genes, though perhaps not to all that will be involved in the stabilized, or buffered, time trajectory, since some may be brought into activity by the operation of the earlier active ones. The priming, moreover, has to be able to persist through cell division and be capable of transmission through many generations of cell proliferation. Few concrete suggestions as to the mechanism have yet been made. One is that the primed genes are already producing the ribonucleic acid molecules, called messenger RNA's, which direct protein synthesis in the cell, but that these messengers are in some way inactivated or prevented from activating the protein-synthesizing machinery; this is known as the "masked messenger" hypothesis. Arguments in favour of this hypothesis are, however, circumstantial rather than direct. In some cases, for instance that of the Drosophila imaginal buds, there is direct evidence against it. Another hypothesis, perhaps more attractive, but much vaguer, is that the determination or priming involves the intervention of some of the large amounts of reiterated DNA known to be present in the cells of higher organisms. At the present time, however, biology lacks any convincing theory of determination in terms of gene action.

It appears at first sight that more is known about actual differentiation than initial determination. Actual differentiation must involve the controlled synthesis of particular proteins, coded for by specific genes. Certainly, a great deal is known about the mechanisms that control the action of genes in directing the synthesis of proteins in simple organisms such as viruses and bacteria. It is tempting to suppose that similar systems operate in controlling the synthetic activities of genes in higher organisms. Unfortunately, no single case of an exactly similar controlling system has ever been discovered in higher organisms, in spite of an intense search for it. It may in fact be suggested that until there is a fuller understanding of the mechanism of "priming" genes at the time of determination, there can scarcely be an adequate account of the way in which the activity of these genes is controlled at later stages.

Basic Concepts of Developmental Biology

2

- **Aging**
- **Compartment**
- **Embryology**
- **Embryogenesis**
- **Metamerism**
- **Metamorphosis**
- **Ontogeny**
- **Teratology**
- **Regeneration**

Some of the most important concepts of developmental biology are embryology, aging, embryogenesis, metamerism, metamorphosis, ontogeny, teratology and regeneration. This chapter has been carefully written to provide an easy understanding of these concepts of developmental biology.

Aging

Aging are the progressive physiological changes in an organism that lead to senescence, or a decline of biological functions and of the organism's ability to adapt to metabolic stress.

Aging takes place in a cell, an organ, or the total organism with the passage of time. It is a process that goes on over the entire adult life span of any living thing. Gerontology, the study of the aging process, is devoted to the understanding and control of all factors contributing to the finitude of individual life. It is not concerned exclusively with debility, which looms so large in human experience, but deals with a much wider range of phenomena. Every species has a life history in which the individual life span has an appropriate relationship to the reproductive life span and to the mechanism of reproduction and the course of development. How these relationships evolved is as germane to gerontology as it is to evolutionary biology. It is also important to distinguish between the purely physicochemical processes of aging and the accidental organismic processes of disease and injury that lead to death.

Gerontology, therefore, can be defined as the science of the finitude of life as expressed in the three aspects of longevity, aging, and death, examined in both evolutionary and individual (ontogenetic) perspective. Longevity is the span of life of an organism. Aging is the sequential or progressive change in an organism that leads to an increased risk of debility, disease, and death. Senescence consists of these manifestations of the aging process.

The viability (survival ability) of a population is characterized in two actuarial functions: the survivorship curve and the age-specific death rate, or Gompertz function. The relation of such factors as aging characteristics, constitutional vigour, physical factors, diet, and exposure to disease-causing organisms to the actuarial functions is complex. There is, nevertheless, no substitute for them as measures of the aging process and of the effect of environmental or genetic modifiers.

The age-specific mortality rate is the most informative actuarial function for investigations of the aging process. It was first pointed out by an English actuary, Benjamin Gompertz, in 1825 that the mortality rate increases in geometric progression—i.e., by a constant ratio in successive equal age intervals. Hence, a straight line, known as the Gompertz function, results when death rates are plotted on a logarithmic (ratio) scale. The prevalence of many diseases and disabilities rises in the same geometrical manner as does the mortality rate, important exceptions being some infectious diseases and diseases arising from disturbances of the immunological system. Although the life tables of most species are remarkably similar in form, even closely related species can differ markedly in the relative incidence of the major causes of death.

For humans in industrialized countries, life expectancy has increased significantly. Indeed, at the beginning of the 20th century, life expectancy in those countries was between 30 and 45 years. At the century's close, life expectancy averaged about 67 years, thanks in large part to improvements in health care, nutrition, and standards of living. In the early 21st century, demographic projections suggested that life expectancy for men and women who maintained the healthiest lifestyle patterns would continue to increase. In the first decade of the 21st century in the United States, centenarians—those who live to age 100 or older—were the fastest-growing segment of the population.

Biological Theories of Aging

Aging has many facets. Hence, there are a number of theories, each of which may explain one or more aspects of aging. There is, however, no single theory that explains all of the phenomena of aging.

Genetic Theories

One theory of aging assumes that the life span of a cell or organism is genetically determined—that the genes of an animal contain a "program" that determines its life span, just as eye colour is determined genetically. This theory finds support in the fact that people with parents who have lived long lives are likely to live long themselves. Also, identical twins have life spans more similar in length than do non-twin siblings.

The genetic theory of aging centres on telomeres, which are repeated segments of DNA (deoxyribonucleic acid) occurring at the ends of chromosomes. The number of repeats in a telomere determines the maximum life span of a cell, since each time a cell divides, multiple repeats are lost. Once telomeres have been reduced to a certain size, the cell reaches a crisis point and is prevented from dividing further. As a consequence, the cell dies.

Research has shown that telomeres are vulnerable to genetic factors that alter an organism's rate of aging. In humans, variations in a gene known as TERC (telomerase RNA [ribonucleic acid] component), which encodes an RNA segment of an enzyme known as telomerase, have been associated with reduced telomere length and an increased rate of biological aging. Telomerase normally functions to prevent the overshortening of telomeres, but in the presence of TERC mutations the enzyme's activity is altered. TERC also appears to influence the telomere length that individuals possess from the time of birth. Persons who carry TERC variations are believed to be several years older biologically compared with noncarriers of the same chronological age. This accelerated rate of biological aging is likely also influenced by exposure to environmental factors, such as smoking and obesity, which increase a carrier's susceptibility to the onset of age-related diseases relatively early in adult life.

Mutations of genes that affect telomere length lend support to another genetic theory of aging, which assumes that cell death is the result of "errors" introduced in the formation of key proteins, such as enzymes. Slight differences induced in the transmission of information from DNA molecules of the chromosomes through RNA molecules (the "messenger" substance) to the proper assembly of the large and complex enzyme molecules could result in a molecule of the enzyme that would not "work" properly. This is precisely what happens in the instance of mutations in the *TERC* gene. Such mutations disrupt the normal function of the telomerase enzyme.

As cells grow and divide, a small proportion of them undergo mutation. This change in the genetic code is then reproduced when the cells again divide. The "somatic mutation" theory of aging assumes that aging is due to the gradual accumulation of mutated cells that do not perform normally.

Nongenetic Theories

Other theories of aging focus attention on factors that can influence the expression of a genetically determined "program." These theories all attempt to explain aging in terms of cellular and molecular changes. Actually, age changes are much more marked in the overall performance of an individual than in cellular processes that can be measured. The age decrement in the ability to perform muscular work is much greater than any changes that can be detected in the enzyme activities of the muscles that perform the work. It is possible that aging in an individual is actually due to a breakdown in the control mechanisms that are required in a complex performance. Aging could also be the result of an accumulation in cells of damaging reactive molecules produced as byproducts of day-to-day cellular activities, such as cellular respiration. Other nongenetic theories consider aging as a complex psychosociological process.

Wear-and-tear Theory

The "wear-and-tear" theory assumes that animals and cells, like machines, simply wear out. Animals, however, unlike machines, have some ability to repair themselves, so that this theory does not fit the facts of a biological system. A corollary to the wear-and-tear theory is the presumption that waste products accumulate within cells and interfere with function. The accumulation of highly insoluble particles, known as "age pigments," has been observed in muscle cells in the heart and nerve cells of humans and other animals.

Cross-linking Theory

With increasing age, tendons, skin, and even blood vessels lose elasticity. This is due to the formation of cross-links between or within the molecules of collagen (a fibrous protein) that give elasticity to these tissues. The "cross-linking" theory of aging assumes that similar cross-links form in other biologically important molecules, such as enzymes. These cross-links could alter the structure and shape of the enzyme molecules so that they are unable to carry out their functions in the cell.

Autoimmune Theory

Another theory of aging assumes that immune reactions, normally directed against disease-producing organisms as well as foreign proteins or tissue, begin to attack cells of the individual's own body. In other words, the system that produces antibodies loses its ability to distinguish between "self" and foreign proteins. This "autoimmune" theory of aging is based on clinical rather than on experimental evidence.

Glycation Theory

"Glycation" theory suggests that glucose acts as a mediator of aging. Glycation, in which simple sugars (e.g., glucose) bind to molecules such as proteins and lipids, has a profound cumulative effect during life. Such effects may be similar to the elevated glucose levels and shorter life spans observed in diabetic humans.

Oxidative Damage Theory

Reactions that take place within cells can result in the oxidation of proteins and other cellular molecules. Oxidation entails the loss of electrons from these molecules, causing them to become unstable and highly reactive and leading to their eventual reaction with and damage of cell components such as membranes. Such reactive molecules are known as free radicals—any atom or molecule that has a single unpaired electron in an outer shell.

Oxidative damage (oxidative stress) accumulates with age, and this has given rise to the free radical theory of aging, which is concerned in particular with molecules known as reactive oxygen species (ROS). This theory was first proposed in the 1950s by American gerontologist Denham Harman and was supported in part by evidence that antioxidant proteins, which neutralize free radicals, are more abundant in aging cells, indicating a response to oxidative stress.

The initial free radical theory of aging was later extended to include ROS derived from cellular organelles known as mitochondria, which are the primary sites of energy production in most eukaryotic organisms (eukaryotic cells are cells with clearly defined nuclei). The mitochondrial theory of aging was based on the idea that there exists within mitochondria a vicious oxidation cycle, in which the mutation of mitochondrial DNA impairs the function of proteins in the organelle's respiration machinery, thereby enhancing the production of DNA-damaging oxygen radicals. This in turn results in the accumulation of mutations in mitochondrial DNA and a bioenergetic impairment, characterized by the failure of mitochondria to produce sufficient energy for cells to carry out their daily activities, which leads to tissue dysfunction and degeneration.

A similar mitochondrial theory of aging proposes a mechanism in which electrons leaking from the electron transport chain (ETC), the central component of the organelle's respiration machinery, produce ROS and then damage ETC proteins and mitochondrial DNA, leading to further increases in intracellular ROS levels and a decline in mitochondrial function.

Another consideration is the molecular inflammatory theory of aging, whereby the activation of redox- (oxidation-reduction-) sensitive transcription factors (molecules that control gene activity) by age-related oxidative stress causes increased expression of proinflammatory genes, leading to inflammation in various tissues. This inflammatory cascade is exaggerated during aging and has been linked to many age-associated pathologies, including cancer, cardiovascular disease, arthritis, and several neurodegenerative diseases.

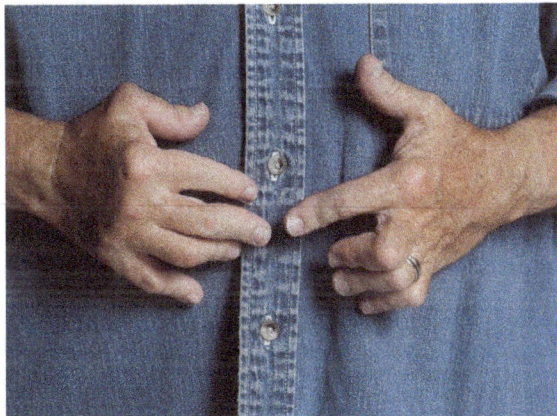

A 65-year-old man with advanced rheumatoid arthritis.

Mammals under calorie restriction produce fewer ROS and age slower. Such effects of calorie restriction have been attributed to its ability to lower the steady state of oxidative stress, slow the accumulation of age-associated oxidative damage, and increase metabolic efficiency.

A common phenomenon in all of the aforementioned theories is that ROS serve as a contributing factor to many age-associated diseases.

Psychosociological Theory

In addition to theories of aging based on molecules and cells, there also exists a "psychosociological" theory of aging. As people grow older, their behaviour changes, their social interactions change, and the activities in which they engage change. The psychosociological theory of aging can be divided roughly into four component theories: disengagement, activity, life-course, and continuity theories. Disengagement theory is based on hampered relationships between a person and other members of society. Activity theory emphasizes the importance of ongoing social activity and suggests that a person's self-concept (self-perspective) is related to the roles held by that person. Life-course theory is based on the developmental stages proposed by German-born American psychoanalyst Erik H. Erikson. According to Erikson's stages, maturity is a process that continues into old age, and in each stage the individual encounters new psychosocial demands. Continuity theory states that older adults try to preserve and maintain internal and external characteristics (e.g., values, personality, preferences, and behaviour patterns) throughout life, despite changes in their health or life circumstances.

Natural History of Aging

Reproduction and Aging

Reproduction is an all-important function of an organism's life history, and all other vital processes, including senescence and death, are shaped to serve it. The distinction between semelparous and iteroparous modes of reproduction is important for an understanding of biological aging. Semelparous organisms reproduce by a single reproductive act. Annual and biennial plants are semelparous, as are many insects and a few vertebrates, notably salmon and eels. Iteroparous organisms, on the other hand, reproduce recurrently over a reproductive span that usually covers a major part of the total life span.

Semelparity; wheat. Cereal crops such as wheat are semelparous, meaning that they die after their first reproduction.

In semelparous forms, reproduction takes place near the end of the life span, after which there ensues a rapid senescence that quickly leads to the death of the organism. In plants the senescent phase is usually an integral part of the reproductive process and essential for its completion. The dispersal of seeds, for example, is accomplished by processes—including ripening and falling (abscission) of fruits and drying of seed pods—that are inseparable from the overall senescence process. Moreover, the onset of plant senescence is invariably initiated by the changing levels of hormones, which are under systemic or environmental control. If, for example, the hormone auxin is prevented, by experimental means, from influencing the plant, the plant lives longer than normal and undergoes an atypical prolonged pattern of senescent change.

Useful inferences can be drawn from the study of the aging processes of insects that display two distinct kinds of adaptive coloration: the procryptic, in which the patterns and colours afford the insect concealment in its native habitat; and the aposematic, in which the vivid markings serve as a warning that the insect is poisonous or bad-tasting. The two adaptation patterns have different optimal species-survival strategies: the procryptics die out as quickly as possible after completing reproduction, thus reducing the opportunity for predators to learn how to detect them; the aposematics have longer post-reproductive survival, thus increasing their opportunity to condition predators. Both adaptations are found in the family of saturniid moths, and it has been shown that the duration of their post-reproductive survival is governed by an enzyme system that controls the fraction of time spent in flight: procryptics fly more, exhaust themselves, and die quickly; aposematics fly less, conserve their energies, and live longer.

Aposematic coloration; tiger moth.

The adaptive coloration patterns of insects represent different survival strategies. For example, tiger moths have aposematic, or warning, coloration, which is associated with prolonged post-reproductive survival, increasing their opportunity to condition predators to their warning strategy.

These examples indicate that in semelparous forms, in which full vigour and function are required until virtually the end of life, senescence has an onset closely coupled with the completion of the reproductive process and is governed by relatively simple enzymatic mechanisms that can be modified by natural selection. Such specific, genetically controlled senescence processes are instances of programmed life termination.

The iteroparous forms include most vertebrates, most of the longer-lived insects, crustaceans and spiders, cephalopod and gastropod mollusks, and perennial plants. In contrast to semelparous forms, iteroparous organisms need not survive to the end of their reproductive phase in order to reproduce successfully, and the average fraction of the reproductive span survived varies widely between groups: small rodents and birds in the wild survive on the average only 10 percent to 20 percent of their potential reproductive lifetimes; whales, elephants, apes, and other large mammals in the wild, on the other hand, live through 50 percent or more of their reproductive spans, and a few survive beyond reproductive age. In iteroparous forms the onset of senescence is gradual, with no evidence of specific systemic or environmental initiating mechanisms. Senescence manifests itself early as a decline in reproductive performance. In species that grow to a fixed body size, decline of reproductive capacity begins quite early and accelerates with increasing age. In large egg-laying reptiles, which attain sexual maturity while relatively small in size and continue to grow during a long reproductive span, the number of eggs laid per year increases with age and body size but eventually levels off and declines. The reproductive span in such cases is shorter than the life span.

These comparisons illustrate the influence exerted by factors of population dynamics on the evolution of reproductive and bodily (somatic) senescence. The proportional contribution of an individual to the rate of increase of the iteroparous population obviously diminishes as the number of its living progeny increases. In addition, the individual's reproductive capacity diminishes with age. These facts imply that there is an optimum number of litters per lifetime. Whether or not these

2

influences of population dynamics lead to the evolution of adaptive senescence patterns has long been debated by gerontologists but has not yet been investigated definitively.

There is some evidence that calorie restriction delays reproductive senescence, which can be at least partially explained by the beneficial effects on the hypothalamus and pituitary gland to enhance the secretion of luteinizing hormone, which helps regulate the activity of the gonads, or sex glands.

Species Differences in Longevity and Aging

There are large differences in life span between some species of animals. The taxonomic stratification of longevity can be seen among the mammals. Primates, generally, are the longest-lived group, although some small prosimians and New World monkeys have relatively short life spans. The murid (mouselike) rodents are short-lived; the sciurid (squirrel-like) rodents, however, can reach ages two to three times longer than the murids.

Squirrel; longevity among mammals. There exist large differences in life span among mammals. In rodents, for example, sciurids such as squirrels may live two to three times longer than murids (mouselike rodents).

Three traits have independent correlations with life span: brain weight, body weight, and resting metabolic rate. The dependence of life span on these traits can be expressed in the form of an equation: $L = 5.5E^{0.54}S^{-0.34}M^{-0.42}$. Mammalian life span ($L$) in months relates to brain weight (E) and body weight (S) in grams and to metabolic rate (M) in calories per gram per hour. The positive exponent for E (0.54) indicates that longevity of mammals has a strong positive association with brain size, independent of body size or metabolic rate. The negative coefficient for metabolic rate implies that life span decreases as the rate of living increases, if brain and body weight are held constant. The negative partial coefficient for body weight indicates that the tendency for large animals to be longer-lived results not from body size but rather from the high positive correlation of body weight with brain weight and its negative correlation with metabolic rate. The same kind of relation of L to E, S, and M holds for birds, but there is a tendency for birds to be longer-lived than mammals of comparable brain and body size despite their higher body temperatures and metabolic rates. The larger reptiles have life spans exceeding those of mammals of comparable size, but their rates of metabolism are about 10 times lower, so that their total lifetime energy expenditures are lower than those for mammals. The more highly cephalized animals (i.e., those with higher brain weight), especially the primates, have greater lifetime energy outputs. The total lifetime energy output per gram of tissue is about 1,200,000 calories for humans and 400,000 calories for domestic animals such as cats and dogs.

The above relations hold for the homeothermic mammals, those with nearly constant body temperature. The heterothermic mammals, which are able to enter daily torpor, or seasonal hibernation, thereby reduce their metabolic rates more than 10-fold. The insectivorous bats of temperate latitudes are the most dramatic example: although they have life spans in excess of 20 years, almost 80 percent of that time is spent in deep torpor, and, as a result, their lifetime energy expenditures are no greater than are those of other small mammals.

The longevities of arthropod species extend from a few days to several decades. The extremely short-lived insects have a brief single reproductive phase; the longer-lived spiders and crustaceans are iteroparous, with annual reproductive cycles.

The Inheritance of Longevity

The inheritance of longevity in animal populations such as fruit flies and mice is determined by comparing the life tables of numerous inbred populations and some of their hybrids. The longevity of sample populations has been measured for more than 40 inbred strains of mice. Two experiments concur in finding that about 30 percent of longevity variation in female mice is genetically determined, whereas the heritability in male mice is about 20 percent. These values are comparable to the heritabilities of some physiological performances in domestic animals, such as lifetime egg or milk production. The slope of the Gompertz function line indicates the rate of actuarial aging. The differences in longevity between species are the result primarily of differences in the rate of aging and are therefore expressed in differences in the slope of the Gompertz function.

Comparison of life tables between mouse strains of a single species indicates that the strain differences result primarily from differences in age-independent hardiness factors. If strains differ in hardiness, the less hardy have death rates higher by a constant multiple at all ages, as shown by the parallel Gompertz functions. It is frequently found that the first-generation (F_1) hybrids of two inbred strains live longer than either parent. There has been no direct comparison of hybrid and inbred mice with regard to the rates of their biochemical aging processes, but life table comparisons indicate that hybrid vigour (heterosis) is an increase of age-independent vigour and not a decrease in the rate of aging.

Much of the variation in survival time between mouse strains is attributable to differences in inherited susceptibility to specific diseases. An important task of gerontology is to determine the extent of such genetic influences on aging.

The inheritance of longevity in humans is more difficult to investigate because length of life is influenced by socioeconomic and other environmental factors that generate spurious correlations between close relatives. A number of studies have been published, most of them pointing to some degree of heritability with regard to length of life or susceptibility to major diseases, such as cancer and heart disease. Although there is disagreement about the degree of heritability of longevity in humans, the evidence for genetic transmission of susceptibility to coronary heart disease and related diseases is strong, as is the evidence that monozygotic (genetically identical) twins tend to have more similar life spans than do like-sex dizygotic (genetically different, fraternal) twins.

Senescence in Mammals

Changes in Body Composition, Metabolism and Activity

The lean body mass, consisting of the skeletal muscles and all other cellular tissues, decreases steadily after physical maturity until, in extreme old age, it may be reduced to two-thirds its value in young adults. Body weight, however, usually increases with age, because stored fat and body water increases in excess of the loss of lean body mass. The relative amount of extracellular fluid increases with age during adult life, after decreasing steadily throughout fetal and postnatal development. Despite appearances, therefore, all tissues, even the skin, become more laden with water as a consequence of aging. The steady loss of voluntary (striated) muscle tissue mass throughout adult life depends somewhat on the pattern of physical activity. Evidence indicates that a large part of the loss of muscle mass with age is the result of disuse and atrophy rather than loss of muscle fibres.

Human aging. Senior citizens participating in an exercise
class for the elderly. Exercise can help slow the loss of
muscle mass associated with aging and senescence.

The decrease of lean body mass is accompanied by a decrease in the level of overall metabolic activity. Basal metabolism is greatest during the period of most rapid mass growth. It then declines rapidly until physical maturity is reached and more slowly thereafter. In the rat the slow phase of decrease amounts to about 20 percent over a three-year period. The interior body temperature is maintained, despite lower heat production, by decreased blood flow through the skin with a consequent decrease of heat loss. The "cooling of the blood" with age, therefore, does not occur in the degree that might be inferred from the decrease in skin temperature. The amount of voluntary physical activity, such as running in an exercise wheel, typically decreases with age but varies considerably between individual animals.

In humans, overall aging-related changes in metabolism that result in increased fat deposition and reduced muscle mass can lead to an increased likelihood of developing metabolic diseases such as type II diabetes mellitus, hyperlipidemia (elevated blood levels of lipids), arteriosclerosis (hardening of the arteries), and hypertension (high blood pressure). In some persons, these conditions may occur simultaneously, giving rise to a condition known as metabolic syndrome.

Also in humans, a substance known as ghrelin, which is produced and secreted mainly by the gastric mucosa and which stimulates food intake, decreases with age. Circulating ghrelin levels decline during aging because of impaired function of the gastric mucosa. This decline is thought to be related to the loss of appetite and anorexia often observed in aged subjects.

Changes in Structural Tissues

The structural integrity of the vertebrate organism depends on two kinds of fibrous protein molecules, collagen and elastin. Collagen, which constitutes almost one-third of the body protein, is found in skin, bone, and tendons. When first synthesized by cells called fibroblasts, collagen is in a fragile and soluble form (tropocollagen). In time this soluble collagen changes to a more stable, insoluble form that can persist in tissues for most of an animal's life. The rate of collagen synthesis is high in youth and declines throughout life, so that the ratio of insoluble to soluble collagen increases with age. Insoluble collagen then builds up with age as a result of synthesis exceeding removal, much like another fibrous tissue, the crystalline lens of the eye. With increasing age, the number of cross-linkages within and between collagen molecules increases, leading to crystallinity and rigidity, which are reflected in a general body stiffness. There is also a decrease in the relative amount of a mucopolysaccharide (i.e., the combination of a protein and a carbohydrate) ground substance; a measure of this, the hexosamine–collagen ratio, has been investigated as an index of individual differences in the rate of aging. An important consequence of these changes is decreased permeability of the tissues to dissolved nutrients, hormones, and antibody molecules.

The rate of aging of collagen is related to the overall metabolic activity of the animal; rats kept on low-calorie diets have more youthful collagen than fully nourished rats of the same age.

Elastin is the molecule responsible for the elasticity of blood vessel walls. With age, progressive loss of elasticity of vessels occurs, presumably because of fragmentation of the elastin molecule.

The cross-linkage of collagen is chemically similar to the cross-linkages that occur in skins when they are tanned to leather. This similarity has stimulated proposals that chemicals that inhibit cross-linkage in tanning will retard aging.

Tissue Cell Loss and Replacement

The tissues of the body fall into two groups, according to whether or not there is continuous renewal of tissue cells. At one extreme are nonrenewal tissues such as nerves and voluntary muscles, in which few new cells are formed (at least in mammals) after a certain stage of growth. In renewal tissues such as the intestinal epithelium and the blood, on the other hand, some cell types live only one or a few days and must be replaced hundreds of times in the life span of even a short-lived animal such as the rat. Between these limits lie many organs, such as liver, skin, and endocrine organs, that have cells that are replaced over periods ranging from a few weeks to several years in humans.

A peripheral nerve is a convenient object to study because the total number of fibres in the nerve trunk can be counted. This has been done for the cervical and thoracic spinal nerve roots of the rat, the cat, and humans. In the ventral and dorsal spinal roots of humans, the number of nerve fibres decreases about 20 percent from age 30 to age 90. In the cat, the rat, and the mouse, however, the data do not consistently indicate a decrease of number of spinal root fibres with age. In humans the number of olfactory nerve fibres, which serve the sense of smell, decreases by age 90 to about 25 percent of the number present at birth, and the number of optic nerve fibres, serving vision, decreases at a nearly comparable rate.

There is a striking decrease in the number of living cells in the cerebral cortex of the brain of humans with age. The cerebellar cortex of the rat and human is about as susceptible to age deterioration as is the cerebral cortex. Other parts of the brain are not so obviously marked by aging.

There is, in short, a tendency for the higher and more recently evolved levels of the nervous system to undergo more severe aging loss than do other regions, such as the brainstem and spinal cord. It is not yet known how much of the loss of brain cells results from conditions within the brain itself and how much results from extrinsic causes, such as deterioration of the blood circulation. The nutrition and maintenance of nerve cells, or neurons, in the central nervous system depends to a considerable extent on neuroglia, small cells that surround the neurons. The absolute number of these cells apparently does not decrease with age, but some of the microscopic changes seen in the neurons of old persons are similar to the changes produced by starvation or physical exhaustion.

It has been shown that after an attack of measles, the virus remains in the host's body for the remainder of life and infrequently gives rise to a rapidly progressing degeneration of the cerebral cortex. This virus or other inapparent viruses may also be responsible for the individual differences in onset of senility in humans.

The renewal tissues are typically made up of a population of proliferative cells, which retain the capability for division, and a population of mature cells, produced by the proliferative cells and with limited life spans. The production of cells must balance the steady loss and also compensate quickly for unusual losses caused by injury or disease, so each renewal tissue has one or more channels of feedback control to adjust production to demand. Aging of renewal tissues is expressed in several ways, including decrease in the number of proliferative cells, decrease in the rate of cell division, and decrease in responsiveness to feedback signals. Changes of these factors in the blood-forming tissues of the mouse are small, yet the blood-forming tissues do suffer an aging deficit, for the ability to respond to extreme or repeated demand is significantly reduced in older mice.

The intact skin has a cell turnover time of several weeks, with the capability, shared by all renewal tissues, of temporarily increasing the rate of cell production by a large factor in response to injury. The rate of wound healing decreases with age, rapidly at first and more slowly as age increases.

One of the most regular and striking aging processes is the decrease in the ability to focus on both close and distant objects. This loss in visual accommodation is the result in part of a weakening of the ciliary muscle of the eye and of a decrease in the flexibility of the lens. A further contributing factor, however, is that the lens continues to grow throughout life at a rate that diminishes with age. This growth is the result of continuous division of epithelial cells near an imaginary midline of the lens, giving rise to fresh cells that differentiate into the precisely aligned lens fibres. Once formed, the fibres remain permanently in place.

An important feature of the renewal mechanism is the stem cell. These cells, which may normally continue to divide at a low rate throughout life, under conditions of increased demand enter a compensatory proliferative phase during which they divide rapidly. Blood-forming tissue has a stem cell population that responds to injury readily in youth, but its capacity diminishes with age. The increased incidence of anemia in old age and the reduced capacity to respond to blood loss have been attributed to depletion of the blood-forming stem cells. Stem cell populations have not been identified with certainty in other proliferative tissues. The intestinal mucosa, in particular, has a high cell-division rate without any clear indication of a reserve population of stem cells.

Mammalian Cell Cultures

Dividing cells from various mammalian tissues can be grown in vitro (outside the body) under careful laboratory control. Various lines of cancer cells have been grown in continuous culture for many decades. In the early period of tissue-culture technology it was claimed that certain chicken cells (fibroblasts) had been maintained in culture for 20 years. This led to the belief that dividing cells were potentially immortal and focused interest on nondividing cells as the seat of the aging process. However, it has since been established that a population (clone) of fibroblasts has a finite life history in culture. It has a period of healthy growth, during which it can be transferred, or "split," several dozen times, indicating that the cells have undergone more than that number of generations. The cultures, however, go into a senescent phase and die out, usually before the 50th transfer. Occasionally, the chromosomes in a cell in the culture undergo a mutation (change) that results in a loss of a growth-limiting factor, leading to the establishment of a subclone capable of indefinite growth. This happens fairly often in cultures of mouse cell strains but only rarely in cultures of human cells. Such mutations usually involve chromosomal rearrangements or changes in the number of chromosomes.

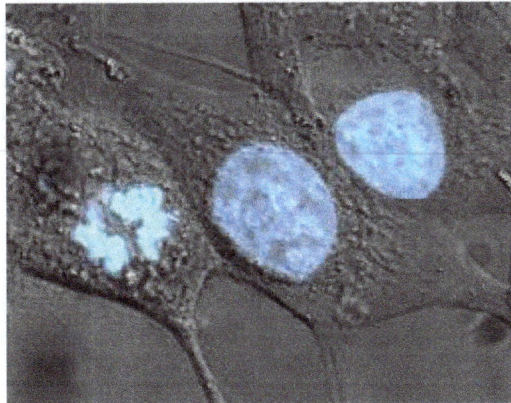

Cell culture. Cultured HeLa cells (cancerous cervical cells) stained with fluorescent Hoechst dye, which turns their nuclei blue.

Thus, dividing mammalian cells with a normal chromosomal complement have a limited growth potential. The capacity for indefinite growth shown by cancer cells and transformed cells is the result of the loss of a growth-limiting factor, such as the loss of control over cell life span normally exerted by telomeres. The number of transfers that cell strains can undergo decreases as the age of the donor increases, in a way reminiscent of the decreased turnover rate of fibroblasts in living chickens and of the decreased rate of wound healing with age.

Changes in Tissue and Cell Morphology

There are numerous instances of tissue changes with age. The atrophy of tissues of moderate degree is usual. The shrinkage of the thymus is especially striking and important in view of its role in immunological defense. The diminution of cellular tissue and replacement by fatty or connective tissue is prominent in bone marrow and skin. In the kidney, entire secretory structures (nephrons) are lost. The secretory cells of the pancreas, thyroid, and similar organs decrease in numbers.

In addition, connective tissues change, becoming increasingly stiff. This makes the organs, blood vessels, and airways more rigid. Cell membranes also change, and many tissues become less efficient in exchanging carbon dioxide and other wastes for oxygen and nutrients. Some tissues may become nodular or more rigid.

An important age change is the accumulation of pigments and inert—possibly deleterious—materials within and between cells. The pigment lipofuscin accumulates within cells of the heart, brain, eye, and other tissues. In humans it is not detectable at a young age, but particularly in the heart it increases to make up a small percentage of the cell volume by old age. Amyloid, an insoluble protein-carbohydrate complex, increases in tissues as a result of aging. It is presumably a product of autoimmune reactions, immune reactions misdirected against the organism itself. In an extreme case of a rare autoimmune disease, amyloidosis, particular organs are virtually choked with amyloid substance.

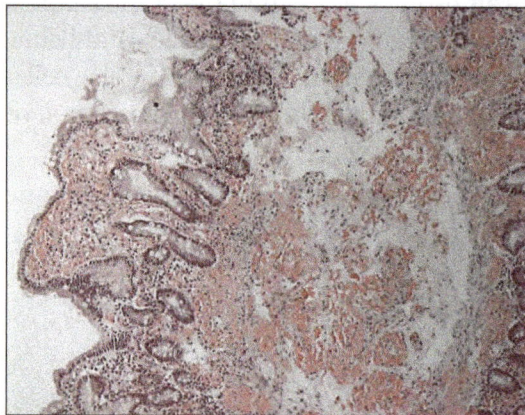

Amyloidosis. Duodenum with amyloid deposition
(bright pink stain) in lamina propria.

Trace metals also accumulate in various tissues with age, and, although the amounts are very small, certain metals can poison enzyme systems and stimulate mutations, which may lead to cancer.

Aging at the Molecular and Cellular Levels

Aging of Genetic Information Systems

The physical basis of aging is either the cumulative loss and disorganization of important large molecules (e.g., proteins and nucleic acids) of the body or the accumulation of abnormal products in cells or tissues. A major effort in aging research has been focused on two objectives: to characterize the molecular disruptions of aging and to determine if one particular kind is primarily responsible for the observed rate and course of senescence; and to identify the chemical or physical reactions responsible for the age-related degradation of large molecules that have either informational or structural roles.

The working molecules of the body, such as enzymes and contractile proteins, which have short turnover times, are not thought to be sites of primary aging damage. Rather, the deoxyribonucleic acid (DNA) molecules of the chromosomes appear to be potential sites of primary damage, because damage to DNA corrupts the genetic message on which the development and function of the organism depend. Damage at a single point in the DNA molecule can be followed by the synthesis of an incorrect protein molecule, which may result in the malfunction or death of the host cell or even of the entire organism. Attention therefore has been given to the somatic mutation hypothesis, which asserts that aging is the result of an accumulation of mutations in the DNA of somatic (body) cells. Aneuploidy, the occurrence of cells with more or less than the correct (euploid) complement of chromosomes, is especially common. The frequency of aneuploid cells in human females increases from 3 percent at age 10 to 13 percent at age 70. Each DNA molecule consists of two complementary strands coiled around each other in a double helix configuration. Evidence indicates that breaks of the individual strands occur with a higher frequency than was once suspected and that virtually all such breaks are repaired by an enzymatic mechanism that destroys the damaged region and then resynthesizes the excised portion, using the corresponding segment of the complementary strand as a model. The mutation rate for a species is therefore governed more by the competence of its repair mechanism than by the rate at which breaks occur. This may help to explain why the mutation rates of different species are roughly proportional to their generation times and justifies research to determine whether the enzymatic mechanisms involved are accessible to control.

There are, however, serious objections to the somatic mutation theory. The wasp *Habrobracon* is an insect that reproduces parthenogenetically (i.e., without the need of sperm to fertilize the egg). It is possible to obtain individuals with either a diploid, or paired, set of chromosomes, as in most higher organisms, or a haploid, single, set. Any gene mutation in a haploid cell at an essential position would result in loss of a vital process and impairment or death of the cell. In a diploid cell a serious mutation is often compensated for by the complementary gene and the cell can carry on its vital functions. Experiments have shown that haploid wasps live about as long as diploids, implying either that mutations are not a quantitatively important factor in aging or that parthenogenetic species have compensated for the vulnerability of their haploids by developing an increased effectiveness of DNA repair.

Chromosomes can be separated into DNA and protein molecules, but with increasing difficulty in older cells. The isolated DNA of old animals, however, does not differ from that of the young. Although most of the DNA in a given cell at a given time is repressed (i.e., blocked from functioning), it is more repressed in old animals.

Aging of the Immune System

Another important molecular information system of the body is the immune system, part of which, the thymus-dependent subsystem, is specialized for defense against invading microorganisms and for detecting and removing body cells that have changed in such ways that they are no longer recognized by the body as part of its own substance, leading to the autoimmune reactions. The immune system has been implicated in the body's defenses against cancer. Cancerous growths (neoplasms) are thought to arise from single cells that undergo a drastic transformation as a result of either a genetic mutation or the activation of a latent (hidden) virus that may have been transmitted genetically from parent to offspring. The control of cancer susceptibility by genetically governed defense mechanisms has been indicated by the breeding of high and low cancer susceptibility in mice. There is a growing body of evidence that the thymus-dependent immune system is instrumental in repressing the development of cancer.

One piece of evidence is that the immunosuppressive procedures of organ transplantation are often followed by a greatly increased incidence of neoplasms. The thymus-dependent system can itself, however, give rise to age-related autoimmune disease, in which the immune system perceives normal body tissue as foreign and attacks it with antibodies. The initial step in these diseases is considered to be a somatic mutation in a single cell of the immune system. Such considerations are the basis of several immune theories of aging, which seek to explain the phenomena of senescence in terms of mutations in the immune system.

Aging of Neural and Endocrine Systems

Aging of the brain entails both degeneration and neuroplasticity. Neurons atrophy and die, and blood flow to the brain decreases. The latter can result in reduced oxygen delivery to tissues, including the eyes and brain. The ability of the eye to dark-adapt (i.e., increase its sensitivity at low light levels) decreases with age, but part of that decrease can be restored by breathing pure oxygen. Various mental processes in elderly people are also found to be improved by breathing oxygen. The establishment of a memory trace (connections in the brain that are associated with memory) involves the synthesis of protein. Any slowed induction of protein synthesis, as from lower oxygen intake, with age could be a factor in the deficits of learning and memory in the elderly. At the same time that neurons are degenerating, however, the aging brain also forms new synapses (connections between neurons), which helps to compensate for the neuronal loss.

A general characteristic of aging of the endocrine system is that the cells that once responded vigorously to hormones become less responsive. A normal chemical in cells, cyclic adenosine monophosphate (AMP), is thought to be a transmitter of hormonal information across cell membranes. It may be possible to identify the specific sites in the membrane or the cell interior at which communication breaks down.

Because the pituitary gland connects the nervous and endocrine systems, its aging affects both systems. In the pituitary, aging attenuates the response of the gland to growth hormone-releasing hormone. This in turn causes a decrease in the release of growth hormone, which subsequently affects the overall rate and efficiency of metabolic processes.

External Environmental Agents

Ionizing Radiation

The shortening of life caused by ionizing radiation (e.g., X-rays) has been determined for many species, including mice, rats, hamsters, guinea pigs, and dogs. The occurrence of some diseases, such as leukemia, may increase disproportionately after irradiation, with the degree of increase influenced by age and sex.

The permanent nature of radiation damage is shown by the comparison of life spans of irradiated and control populations. An irradiated population dies out like a chronologically older unirradiated population. Members of a population given a single dose of X-rays or gamma rays in early adult life die of the same diseases that afflict the unirradiated control population, but they die months or even years earlier.

Continuous irradiation throughout life at low dose rates (daily doses from one-thousandth to one-tenth the dose that would kill immediately) speeds the mortality process. Studies of animals and of cells grown in culture suggest that large doses of radiation kill by producing deleterious rearrangements of chromosomes in the proliferative cell population. Such aberrations also increase with age, but they seem to be less important in the natural aging process. At low radiation doses, chromosome aberrations become relatively less important than other effects, and the primary radiation damage in these conditions may bear a closer relation to the aging lesion.

Natural radioactivity in the body, arising mostly from radioactive potassium and radium, and natural background irradiation, from Earth and from cosmic rays, are not major contributors to the aging process, even in the long-lived human species. They are responsible, however, for a small percentage of cancer incidence. Although the dose to the body from medical radiations is a fraction of the background level and the radiation from nuclear weapon test fallout is less than 1 percent of the background, both sources contribute to cancer induction in proportion to their amounts.

Temperature

Flour beetles, fruit flies, fishes, and other poikilothermic (temperature-variable) organisms live longer at the lower range of environmental temperature. These observations led to the rate-of-living hypothesis, which, simply stated, holds that an organism's life span is dependent on some critical substance that is exhausted more rapidly at higher temperature. Careful analysis of the data on temperature–longevity relations shows, however, that the rate-of-living hypothesis is inadequate in its original form. The most telling evidence comes from experiments in which fruit flies were kept at one temperature for part of their lives and at another temperature for the remainder. The results are not consistent with the rate-of-living hypothesis, but no satisfactory theory has appeared as yet to take its place. An important factor that has not yet been adequately taken into account is the relation of metabolic efficiency to temperature. The energy cost of the biosynthetic processes studied has been discovered to be minimal at an intermediate temperature in the range to which the species is adapted and to increase at higher or lower temperatures. A related phenomenon holds for longevity; the number of calories expended by fruit flies per lifetime is maximal at an intermediate temperature, so the rate of aging per calorie is minimal at that temperature.

There is a question of the degree to which aging occurs as a result of heat destruction (thermal denaturation) of proteins. Thermal denaturation is predominately a disruption of the folding of molecules, which requires the breaking of numbers of low-energy bonds. It seems not to be a strong contributing factor to aging. There is still the possibility that rare events, such as mutations, may arise to a significant degree from thermal denaturation.

Research has suggested that humans might live longer if their core body temperatures were lower, since in shorter-lived species there is a relationship between high metabolism, which increases core temperature, and short life span. In a study of mice engineered to have a lower-than-normal core body temperature, a reduction of about 0.5 °C (0.9 °F) was associated with a roughly 20 percent increase in life span.

Physical Wear of Non-renewable Structures

One of an animal's most important assets is its chewing apparatus, including jaws and teeth. Adaptation to tooth rate of wear is especially important for animals that consume large quantities of grass and herbage. Such adaptations include higher tooth crowns (hypsodonty), larger grinding area, and longer tooth growth period. Tooth wear may be limiting for survival in adverse environments, but, on the whole, it is not an important life-limiting characteristic. The same can be said for other external organs subject to physical wear.

Infectious Disease and Nutrition

The populations in poor environments, characterized by high rates of infectious disease and poor nutrition, have higher death rates than populations in good environments at all ages, yet there is no positive evidence that disadvantaged populations experience a higher rate of aging.

Rats kept on diets restricted in calories live longer and have lower cancer incidence than do rats that are allowed to eat at will. Maximum longevity, however, is achieved at a nutritional level that keeps the animal sexually immature and below normal weight.

Internal Environment: Consequences of Metabolism

The metabolic activities of organisms produce highly reactive chemicals, including strong oxidizing agents. The internal structure of the cell, however, minimizes the harmful effects of such agents. The critical reactions take place within enclosed structures such as ribosomes, membranes, or mitochondria, and counteractive enzymes such as peroxidases are present in abundance. It is nevertheless likely that low concentrations of those reactive substances can reach vital molecules and contribute to the characteristic rate of aging injury. Experiments in which mice are fed low levels of antioxidants such as butylated hydroxytoluene (BHT) have been encouraging but are still somewhat equivocal.

Membranes are the site of much of the metabolic activity of cells; they provide the barriers that keep incompatible reactions separated. Membrane-bound structures known as lysosomes contain enzymes capable of digesting the cell if released. The stability of cells and organisms is therefore very much bound up with the stability of membranes. A number of drugs, including corticosteroids, salicylates, and antihistamines, act by stabilizing cell membranes against inflammatory stimuli. Some of them are found to prolong life in fruit flies and to prolong survival of cells in vitro. The mode of action of these drugs is connected to substances called prostaglandins, which can alter specific membrane characteristics.

Compartment

Compartments can be simply defined as separate, different, adjacent cell populations, which upon juxtaposition, create a lineage boundary. This boundary prevents cell movement from cells from different lineages across this barrier, restricting them to their compartment. Subdivisions are established by morphogen gradients and maintained by local cell-cell interactions, providing functional units with domains of different regulatory genes, which give rise to distinct fates. Compartment boundaries are found across species. In the hindbrain of vertebrate embryos, rhobomeres are compartments of common lineage outlined by expression of Hox genes. In invertebrates, the wing imaginal disc of *Drosophila* provides an excellent model for the study of compartments. Although other tissues, such as the abdomen, and even other imaginal discs are compartmentalized, much of our understanding of key concepts and molecular mechanisms involved in compartment boundaries has been derived from experimentation in the wing disc of the fruit fly.

Function

By separating different cell populations, the fate of these compartments are highly organized and regulated. In addition, this separation creates a region of specialized cells close to the boundary, which serves as a signaling center for the patterning, polarizing and proliferation of the entire disc. Compartment boundaries establish these organizing centers by providing the source of morphogens that are responsible for the positional information required for development and regeneration. The inability of cell competition to occur across the boundary, indicates that each compartment serves as an autonomous unit of growth. Differences in growth rates and patterns in each compartment, maintain the two lineages separated and each control the precise size of the imaginal discs.

Cell Separation

These two cell populations are kept separate by a mechanism of cell segregation linked to the heritable expression of a selector gene. A selector gene is one that is expressed in one group of cells but not the other, giving the founder cells and their descendants different instructions. Eventually these selector genes become fixed in either an expressed or unexpressed state and are stably inherited to the descendants, specifying the identity of the compartment and preventing these genetically different cell populations from intermixing. Therefore, these selector genes are key for the formation and maintenance of lineage compartments.

Central Dogma

The difference in selector gene activity not only establishes two compartments, but also leads to the formation of a boundary between these two that serves as a source of morphogen gradients. In the central dogma of compartments, first, morphogen gradients position founder compartment cells. Then, active/inactive selector genes give a unique genetic identity to cells within a compartment, instructing their fate and their interactions with the neighboring compartment. Finally, border cells, established by short-range signaling from one compartment to its neighboring compartment emit long-range signals that spread to both compartments to regulate the growth and pattering of the entire tissue.

A/P Boundary

In 1970, by means of clonal analysis, the Anterior-Posterior boundary was identified. The founder cells, found at the border between parasegments 4 and 5 of embryo, are already determined at the early blastoderm stage and defined into the two populations they will generate by stripes of the engrailed gene. The selector gene, engrailed (en), is a key determinant in boundary formation between the anterior and posterior compartments. As the wing imaginal disc expands, posterior, but not anterior cells will express engrailed and maintain this expression state as they expand and form the disc. Engrailed mutant clones of posterior origin will gain anterior affinity and move towards the anterior compartment and intermix with those cells. Within the posterior compartment these clones will sort out and form an ectopicborder where they meet other posterior cells.

Similarly, a clone of anterior cells expressing engrailed will gain posterior identity and create an ectopic boundary where the clone meets other anterior cells in this compartment. In addition,to its cell autonomous role in specifying posterior compartment identity, engrailed also has a non-cell autonomous function in the general growth and patterning of the wing disc, through the activation of signaling pathways such as Hedgehog (Hh) and Decapentaplegic (Dpp). The presence of engrailed in the posterior cells leads to the secretion of the short-range inducer Hh which can cross over to the anterior compartment to activate the long-range morphogen, Dpp. Cells in the posterior compartment produce Hh, but only anterior cells can transduce the signal. Optomotor-blind (omb) is involved in the transcriptional response of Dpp, which is only required in the anterior cells to interpret Hh signaling for boundary formation and maintenance. In addition, Cubitus interruptus (Ci), the signal transducer of the Hh signal, is expressed throughout the anterior compartment, particularly in anterior border cells. In posterior cells engrailed prevents the expression of Ci, such it is only expressed in anterior cells and hence only these cells can respond to Hh signaling by up-regulating the expression of dpp. Loss of engrailed function in posterior cells, results in anterior transformation, where Hh expression is decreased and dpp, ci and patched (ptc) is increased, resulting in the formation of a new A/P boundary, suggesting that en positively regulates hh, while negatively regulating ci, ptc and dpp.

Cell Segregation

To explain how anterior and posterior cells are kept separated, the differential adhesion hypothesis proposes that these two cell populations express different adhesion molecules, producing different affinities for each other that minimize their contact. The selector affinity model proposes that difference in cell affinity between compartments is a result of differential selector gene expression. The presence or absence of selector genes in a given compartment produces compartment-specific adhesion or recognition molecules that are different from those in its counterpart. For example, engrailed expressed in the posterior, but not the anterior, cells provides the differential affinity that keeps these compartments separately. It is also possible that this difference in cell adhesion/affinity is not directly due to en expression, but rather to the ability to receive Hh signaling. Anterior cells, capable of Hh transduction, will express given adhesive molecules that would differ from those present in posterior cells, creating differential affinity that would prevent them from intermixing. This signaling-affinity model is supported by experiments that demonstrate the importance of Hh signaling. Clones mutant for the Smoothened (smo), the gene responsible for transducing Hh signaling, retain anterior-like features, but move into the posterior compartment without any changes in the expression engrailed or invected. This demonstrates that Hh signaling, rather than the absence of en, is what gives cells their compartmental identity. Nonetheless, this signaling-affinity model is incomplete: smo mutant clones of anterior origin that migrate into the posterior compartment, do not completely associate with these cells, but rather form a smooth boundary with these posterior cells. If signaling-affinity were the only factor determining compartment identity, then these clones, which are no longer receiving Hh signaling, would have the same affinity as the other posterior cells in that compartment and be able to intermix with them.

These experiments indicate that although Hh signaling could be having an effect in adhesive properties, this effect is limited to the border cells rather than throughout both compartments. It is also possible that both compartments produce the same cell adhesion molecules, but a difference in its abundance or activity could result in sorting between the two compartments. In vitro, transfected cells with high levels of a given adhesion molecule will segregate from cells that expressing lower levels of this same molecule. Finally, differences in cell bond tension could also play a role in the establishment of the boundary and the separation of the two different cell populations. Experimental data has shown that Myosin-II is up-regulated along both the dorsal-ventral and anterior-posterior boundaries in the imaginal wing disc. The D/V boundary is characterized by the presence of filamentous actin and mutations in Myosin-II heavy chain impairs D/V compartmentalization. Similarly, both F-actin and Myosin-II are increased along the A/P boundary, accompanied by a decrease of Bazooka, which was also observed in the D/V border. The Rho-kinase inhibitor Y-27632, of which Myosin-II is the main target, significantly reduces cell bond tension, suggesting that Myosin-II could be the main effector of this process. In support of the signaling-affinity model, creating an artificial interface between cells with active vs. inactive Hh signaling induces a junctional behavior that aligns the cell bonds of where these opposing cell types meet. Moreover, a 2.5-fold increase in mechanical tension is observed along the A/P boundary, compared to the rest of the tissue. Simulations using a vertex model demonstrate that this increase in cell bond tension is enough to maintain proliferating cell populations in separate compartment boundaries. Parameters used to measure cell bond tension are based cell-cell adhesion and cortical tension input. It has also been suggested that boundary formation is not a result of differential mechanical tension between the two cell populations, but could be a result of the mechanical properties of the boundary itself. The level the adhesion molecule, E-cadherin, was unaltered and the biophysical properties of cells between the two compartments were the same. Changes in cell properties, such as an enlarged apical cross-section area, are only observed in anterior and posterior border cells. Along the boundary, orientation of cell divisions was random and there is no evidence that increased cell death or zones of non-proliferating cells are important for maintaining the A/P or D/V boundary.

Embryology

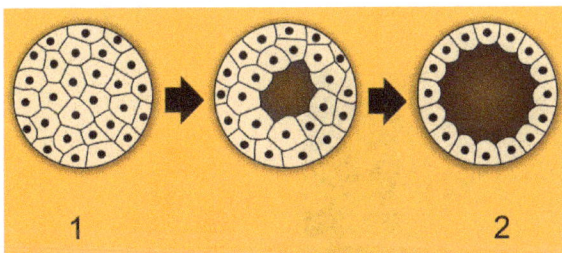

1 - morula, 2 - blastula.

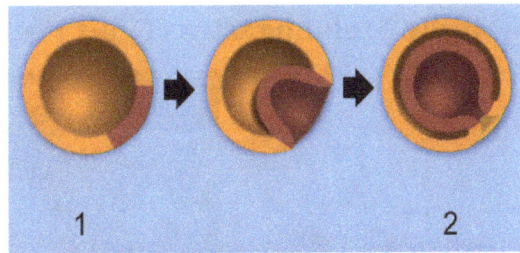

1 - blastula, 2 - gastrula with blastopore;
orange - ectoderm, red - endoderm.

Embryology is the branch of biology that studies the prenatal development of gametes (sex cells), fertilization, and development of embryos and fetuses. Additionally, embryology encompasses the study of congenital disorders that occur before birth, known as teratology.

Embryology has a long history. Aristotle proposed the currently accepted theory of epigenesis, that organisms develop from seed or egg in a sequence of steps. The alternative theory, preformationism, that organisms develop from pre-existing miniature versions of themselves, however, held sway until the 18th century. Modern embryology developed from the work of von Baer, though accurate observations had been made in Italy by anatomists such as Aldrovandi and Leonardo da Vinci in the Renaissance.

Embryonic Development of Animals

After cleavage, the dividing cells, or morula, becomes a hollow ball, or blastula, which develops a hole or pore at one end.

Bilateria

In bilateral animals, the blastula develops in one of two ways that divide the whole animal kingdom into two halves. If in the blastula the first pore (blastopore) becomes the mouth of the animal, it is a protostome; if the first pore becomes the anus then it is a deuterostome. The protostomes include most invertebrate animals, such as insects, worms and molluscs, while the deuterostomes include the vertebrates. In due course, the blastula changes into a more differentiated structure called the gastrula.

The gastrula with its blastopore soon develops three distinct layers of cells (the germ layers) from which all the bodily organs and tissues then develop:

- The innermost layer, or endoderm, give rise to the digestive organs, the gills, lungs or swim bladder if present, and kidneys or nephrites.

- The middle layer, or mesoderm, gives rise to the muscles, skeleton if any, and blood system.

- The outer layer of cells, or ectoderm, gives rise to the nervous system, including the brain, and skin or carapace and hair, bristles, or scales.

Embryos in many species often appear similar to one another in early developmental stages. The reason for this similarity is because species have a shared evolutionary history. These similarities among species are called homologous structures, which are structures that have the same or similar function and mechanism, having evolved from a common ancestor.

Drosophila Melanogaster (Fruit Fly)

Drosophila melanogaster.

Drosophila melanogaster, a fruit fly, is a model organism in biology on which much research into embryology has been done. Before fertilization, the female gamete produces an abundance of mRNA - transcribed from the genes that encode bicoid protein and nanos protein. These mRNA molecules are stored to be used later in what will become the developing embryo. The male and female

Drosophila gametes exhibit anisogamy (differences in morphology and sub-cellular biochemistry). The female gamete is larger than the male gamete because it harbors more cytoplasm and, within the cytoplasm, the female gamete contains an abundance of the mRNA. At fertilization, the male and female gametes fuse (plasmogamy) and then the nucleus of the male gamete fuses with the nucleus of the female gamete (karyogamy). Note that before the gametes' nuclei fuse, they are known as pronuclei. A series of nuclear divisions will occur without cytokinesis (division of the cell) in the zygote to form a multi-nucleated cell (a cell containing multiple nuclei) known as a syncytium. All the nuclei in the syncytium are identical, just as all the nuclei in every somatic cell of any multi-cellular organism are identical in terms of the DNA sequence of the genome. Before the nuclei can differentiate in transcriptional activity, the embryo (syncytium) must be divided into segments. In each segment, a unique set of regulatory proteins will cause specific genes in the nuclei to be transcribed. The resulting combination of proteins will transform clusters of cells into early embryo tissues that will each develop into multiple fetal and adult tissues later in development (note: this happens after each nucleus becomes wrapped with its own cell membrane).

Drosophila melanogaster larvae contained in lab apparatus
to be used for experiments in genetics and embryology.

The process that leads to cell and tissue differentiation is:

Maternal-effect genes - subject to Maternal (cytoplasmic) inheritance.

- Egg-polarity genes establish the Anteroposterior axis.

  Zygotic-effect genes - subject to Mendelian (classical) inheritance.

- Segmentation genes establish 14 segments of the embryo using the anteroposterior axis as a guide:

  ◦ Gap genes establish 3 broad segments of the embryo.

  ◦ Pair-rule genes define 7 segments of the embryo within the confines of the second broad segment that was defined by the gap genes.

  ◦ Segment-polarity genes define another 7 segments by dividing each of the pre-existing 7 segments into anterior and posterior halves.

- Homeotic (homeobox) genes use the 14 segments as pinpoints for specific types of cell differentiation and the histological developments that correspond to each cell type.

Humans

Humans are bilaterals and deuterostomes. In humans, the term embryo refers to the ball of dividing cells from the moment the zygote implants itself in the uterus wall until the end of the eighth week after conception. Beyond the eighth week after conception (tenth week of pregnancy), the developing human is then called a fetus.

Vertebrate and Invertebrate Embryology

Many principles of embryology apply to invertebrates as well as to vertebrates. Therefore, the study of invertebrate embryology has advanced the study of vertebrate embryology. However, there are many differences as well. For example, numerous invertebrate species release a larva before development is complete; at the end of the larval period, an animal for the first time comes to resemble an adult similar to its parent or parents. Although invertebrate embryology is similar in some ways for different invertebrate animals, there are also countless variations. For instance, while spiders proceed directly from egg to adult form, many insects develop through at least one larval stage.

Embryogenesis

Embryonic development also embryogenesis is the process by which the embryo forms and develops. In mammals, the term refers chiefly to early stages of prenatal development, whereas the terms fetus and fetal development describe later stages.

Embryonic development starts with the fertilization of the egg cell (ovum) by a sperm cell, (spermatozoon). Once fertilized, the ovum is referred to as a zygote, a single diploid cell. The zygote undergoes mitotic divisions with no significant growth (a process known as cleavage) and cellular differentiation, leading to development of a multicellular embryo.

Fertilization and the Zygote

The egg cell is generally asymmetric, having an "animal pole" (future ectoderm and mesoderm) and a "vegetal pole" (future endoderm). It is covered with protective envelopes, with different layers. The first envelope – the one in contact with the membrane of the egg – is made of glycoproteins and is known as the vitelline membrane (zona pellucida in mammals). Different taxa show different cellular and acellular envelopes englobing the vitelline membrane.

Fertilization (also known as 'conception', 'fecundation' and 'syngamy') is the fusion of gametes to produce a new organism. In animals, the process involves a sperm fusing with an ovum, which eventually leads to the development of an embryo. Depending on the animal species, the process can occur within the body of the female in internal fertilisation, or outside in the case of external fertilisation. The fertilized egg cell is known as the zygote.

To prevent more than one sperm fertilizing the egg, polyspermy, fast block and slow block to polyspermy are used. Fast block, the membrane potential rapidly depolarizing and then returning to normal, happens immediately after an egg is fertilized by a single sperm. Slow block begins the first few seconds after fertilization and is when the release of calcium causes the cortical reaction, various enzymes releasing from cortical granules in the eggs plasma membrane, to expand and harden the outside membrane, preventing more sperm from entering.

Cleavage and Morula

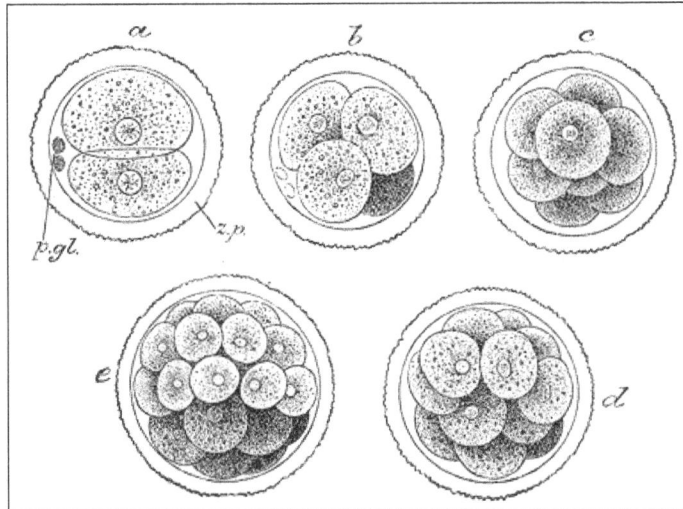

Cell divisions (cleavage).

Cell division with no significant growth, producing a cluster of cells that is the same size as the original zygote, is called cleavage. At least four initial cell divisions occur, resulting in a dense ball of at least sixteen cells called the morula. The different cells derived from cleavage, up to the blastula stage, are called blastomeres. Depending mostly on the amount of yolk in the egg, the cleavage can be holoblastic (total) or meroblastic (partial).

Holoblastic cleavage occurs in animals with little yolk in their eggs, such as humans and other mammals who receive nourishment as embryos from the mother, via the placenta or milk, such as might be secreted from a marsupium. On the other hand, meroblastic cleavage occurs in animals whose eggs have more yolk (i.e. birds and reptiles). Because cleavage is impeded in the vegetal pole, there is an uneven distribution and size of cells, being more numerous and smaller at the animal pole of the zygote.

In holoblastic eggs the first cleavage always occurs along the vegetal-animal axis of the egg, and the second cleavage is perpendicular to the first. From here the spatial arrangement of blastomeres can follow various patterns, due to different planes of cleavage, in various organisms:

Cleavage patterns followed by holoblastic and meroblastic eggs	
Holoblastic	Meroblastic
• Radial (sea urchin, amphioxus)	• Discoidal (fish, birds, reptiles)
• Bilateral (tunicates, amphibians)	• Superficial (insects)
• Spiral (annelids, mollusks)	
• Rotational (mammals, nematodes)	

The end of cleavage is known as midblastula transition and coincides with the onset of zygotic transcription.

In amniotes, the cells of the morula are at first closely aggregated, but soon they become arranged into an outer or peripheral layer, the trophoblast, which does not contribute to the formation of the embryo proper, and an inner cell mass, from which the embryo is developed. Fluid collects between the trophoblast and the greater part of the inner cell-mass, and thus the morula is converted into a vesicle, called the blastodermic vesicle. The inner cell mass remains in contact, however, with the trophoblast at one pole of the ovum; this is named the embryonic pole, since it indicates the location where the future embryo will develop.

Formation of the Blastula

After the 7th cleavage has produced 128 cells, the embryo is called a blastula. The blastula is usually a spherical layer of cells (the blastoderm) surrounding a fluid-filled or yolk-filled cavity (the blastocoel).

Mammals at this stage form a structure called the blastocyst, characterized by an inner cell mass that is distinct from the surrounding blastula. The blastocyst must not be confused with the blastula; even though they are similar in structure, their cells have different fates. In the mouse, primordial germ cells arise from a layer of cells in the inner cell mass of the blastocyst (the epiblast) as a result of extensive genome-wide reprogramming. Reprogramming involves global DNA demethylation facilitated by the DNA base excision repair pathway as well as chromatin reorganization, and results in cellular totipotency.

Before gastrulation, the cells of the trophoblast become differentiated into two strata: The outer stratum forms a syncytium (i.e., a layer of protoplasm studded with nuclei, but showing no evidence of subdivision into cells), termed the syncytiotrophoblast, while the inner layer, the cytotrophoblast or "Layer of Langhans", consists of well-defined cells. As already stated, the cells of the trophoblast do not contribute to the formation of the embryo proper; they form the ectoderm of the chorion and play an important part in the development of the placenta. On the deep surface of the inner cell mass, a layer of flattened cells, called the endoderm, is differentiated and quickly assumes the form of a small sac, called the yolk sac. Spaces appear between the remaining cells of the mass and, by the enlargement and coalescence of these spaces, a cavity called the amniotic cavity is gradually developed. The floor of this cavity is formed by the embryonic disk, which is composed of a layer of prismatic cells, the embryonic ectoderm, derived from the inner cell mass and lying in apposition with the endoderm.

Formation of the Germ Layers

The embryonic disk becomes oval and then pear-shaped, the wider end being directed forward. Near the narrow, posterior end, an opaque streak, called the primitive streak, makes its appearance and extends along the middle of the disk for about one-half of its length; at the anterior end of the streak there is a knob-like thickening termed the primitive node or knot, (known as *Hensen's knot* in birds). A shallow groove, the primitive groove, appears on the surface of the streak, and the anterior end of this groove communicates by means of an aperture, the blastopore, with the yolk sac. The primitive streak is produced by a thickening of the axial part of the ectoderm, the cells of which multiply, grow downward, and blend with those of the subjacent endoderm. From the sides of the primitive streak a third layer of cells, the mesoderm, extends laterally between the ectoderm and endoderm; the caudal end of the primitive streak forms the cloacal membrane. The blastoderm now consists of three layers, named from without inward: ectoderm, mesoderm, and endoderm; each has distinctive characteristics and gives rise to certain tissues of the body. For many mammals, it is sometime during formation of the germ layers that implantation of the embryo in the uterus of the mother occurs.

Formation of the Gastrula

During gastrulation cells migrate to the interior of the blastula, subsequently forming two (in diploblastic animals) or three (triploblastic) germ layers. The embryo during this process is called a gastrula. The germ layers are referred to as the ectoderm, mesoderm and endoderm. In diploblastic animals only the ectoderm and the endoderm are present.

- Among different animals, different combinations of the following processes occur to place the cells in the interior of the embryo:

    ○ Epiboly: Expansion of one cell sheet over other cells.

    ○ Ingression: Migration of individual cells into the embryo (cells move with pseudopods).

    ○ Invagination: Infolding of cell sheet into embryo, forming the mouth, anus, and archenteron.

    ○ Delamination: Splitting or migration of one sheet into two sheets.

    ○ Involution: Inturning of cell sheet over the basal surface of an outer layer.

    ○ Polar proliferation: Cells at the polar ends of the blastula/gastrula proliferate, mostly at the animal pole.

- Other major changes during gastrulation:

    ○ Heavy RNA transcription using embryonic genes; up to this point the RNAs used were maternal (stored in the unfertilized egg).

    ○ Cells start major differentiation processes, losing their totipotentiality.

In most animals, a blastopore is formed at the point where cells are entering the embryo. Two major groups of animals can be distinguished according to the blastopore's fate. In deuterostomes the anus forms from the blastopore, while in protostomes it develops into the mouth.

Formation of the Early Nervous System – Neural Groove, Tube and Notochord

In front of the primitive streak, two longitudinal ridges, caused by a folding up of the ectoderm, make their appearance, one on either side of the middle line formed by the streak. These are named the neural folds; they commence some little distance behind the anterior end of the embryonic disk, where they are continuous with each other, and from there gradually extend backward, one on either side of the anterior end of the primitive streak. Between these folds is a shallow median groove, the neural groove. The groove gradually deepens as the neural folds become elevated, and ultimately the folds meet and coalesce in the middle line and convert the groove into a closed tube, the neural tube or canal, the ectodermal wall of which forms the rudiment of the nervous system. After the coalescence of the neural folds over the anterior end of the primitive streak, the blastopore no longer opens on the surface but into the closed canal of the neural tube, and thus a transitory communication, the neurenteric canal, is established between the neural tube and the primitive digestive tube. The coalescence of the neural folds occurs first in the region of the hind brain, and from there extends forward and backward; toward the end of the third week, the front opening (anterior neuropore) of the tube finally closes at the anterior end of the future brain, and forms a recess that is in contact, for a time, with the overlying ectoderm; the hinder part of the neural groove presents for a time a rhomboidal shape, and to this expanded portion the term sinus rhomboidalis has been applied. Before the neural groove is closed, a ridge of ectodermal cells appears along the prominent margin of each neural fold; this is termed the neural crest or ganglion ridge, and from it the spinal and cranial nerve ganglia and the ganglia of the sympathetic nervous system are developed. By the upward growth of the mesoderm, the neural tube is ultimately separated from the overlying ectoderm.

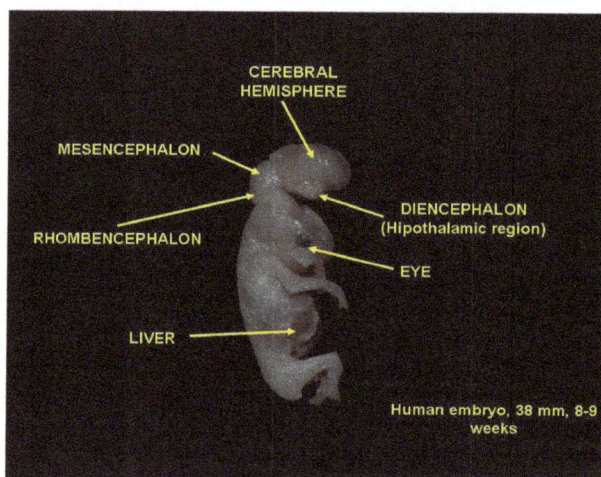

Dissection of human embryo.

The cephalic end of the neural groove exhibits several dilatations that, when the tube is closed, assume the form of the three primary brain vesicles, and correspond, respectively, to the future forebrain (prosencephalon), midbrain (mesencephalon), and hindbrain (rhombencephalon). The walls of the vesicles are developed into the nervous tissue and neuroglia of the brain, and their cavities are modified to form its ventricles. The remainder of the tube forms the spinal cord (medulla spinalis); from its ectodermal wall the nervous and neuroglial elements of the spinal cord are developed, while the cavity persists as the central canal.

Formation of the Early Septum

The extension of the mesoderm takes place throughout the whole of the embryonic and extra-embryonic areas of the ovum, except in certain regions. One of these is seen immediately in front of the neural tube. Here the mesoderm extends forward in the form of two crescentic masses, which meet in the middle line so as to enclose behind them an area that is devoid of mesoderm. Over this area, the ectoderm and endoderm come into direct contact with each other and constitute a thin membrane, the buccopharyngeal membrane, which forms a septum between the primitive mouth and pharynx.

Early Formation of the Heart and other Primitive Structures

In front of the buccopharyngeal area, where the lateral crescents of mesoderm fuse in the middle line, the pericardium is afterward developed, and this region is therefore designated the pericardial area. A second region where the mesoderm is absent, at least for a time, is that immediately in front of the pericardial area. This is termed the proamniotic area, and is the region where the proamnion is developed; in humans, however, it appears that a proamnion is never formed. A third region is at the hind end of the embryo, where the ectoderm and endoderm come into apposition and form the cloacal membrane.

Somitogenesis

Somitogenesis is the process by which somites (primitive segments) are produced. These segmented tissue blocks differentiate into skeletal muscle, vertebrae, and dermis of all vertebrates.

Somitogenesis begins with the formation of somitomeres (whorls of concentric mesoderm) marking the future somites in the presomitic mesoderm (unsegmented paraxial). The presomitic mesoderm gives rise to successive pairs of somites, identical in appearance that differentiate into the same cell types but the structures formed by the cells vary depending upon the anteroposterior (e.g., the thoracic vertebrae have ribs, the lumbar vertebrae do not). Somites have unique positional values along this axis and it is thought that these are specified by the Hox homeotic genes.

Toward the end of the second week after fertilization, transverse segmentation of the paraxial mesoderm begins, and it is converted into a series of well-defined, more or less cubical masses, also known as the somites, which occupy the entire length of the trunk on either side of the middle line from the occipital region of the head. Each segment contains a central cavity (known as a myocoel), which, however, is soon filled with angular and spindle-shape cells. The somites lie immediately under the ectoderm on the lateral aspect of the neural tube and notochord, and are connected to the lateral mesoderm by the intermediate cell mass. Those of the trunk may be arranged in the following groups, viz.: cervical 8, thoracic 12, lumbar 5, sacral 5, and coccygeal from 5 to 8. Those of the occipital region of the head are usually described as being four in number. In mammals, somites of the head can be recognized only in the occipital region, but a study of the lower vertebrates leads to the belief that they are present also in the anterior part of the head and that, altogether, nine segments are represented in the cephalic region.

Organogenesis

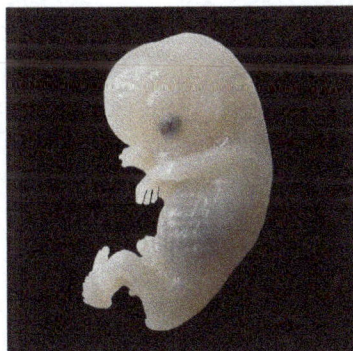

Human embryo, 8-9 weeks, 38 mm.

At some point after the different germ layers are defined, organogenesis begins. The first stage in vertebrates is called neurulation, where the neural plate folds forming the neural tube . Other common organs or structures that arise at this time include the heart and somites, but from now on embryogenesis follows no common pattern among the different taxa of the animal kingdom.

In most animals organogenesis, along with morphogenesis, results in a larva. The hatching of the larva, which must then undergo metamorphosis, marks the end of embryonic development.

Metamerism

Earthworms are a classic example of biological homonymous metamery – the property of repeating body segments with distinct regions.

Metamerism is the phenomenon of having a linear series of body segments fundamentally similar in structure, though not all such structures are entirely alike in any single life form because some of them perform special functions. In animals, metameric segments are referred to as somites or metameres. In plants, they are referred to as metamers or, more concretely, phytomers.

In Animals

In animals, zoologists define metamery as a mesodermal event resulting in serial repetition of unit subdivisions of ectoderm and mesoderm products. Endoderm is not involved in metamery. Segmentation is not the same concept as metamerism: segmentation can be confined only to ectodermally derived tissue, e.g., in the Cestoda tapeworms. Metamerism is far more important biologically since it results in metameres - also called somites - that play a critical role in advanced locomotion.

One can divide metamerism into two main categories:

- Homonymous metamery is a strict serial succession of metameres. It can be grouped into two more classifications known as pseudometamerism and true metamerism. An example of pseudometamerism is in the class Cestoda. The tapeworm is composed of many repeating segments - primarily for reproduction and basic nutrient exchange. Each segment acts independently from the others, which is why it is not considered true metamerism. Another worm, the earthworm in class Annelida, can exemplify true metamerism. In each segment of the worm, a repetition of organs and muscle tissue can be found. What differentiates the Annelids from Cestoda is that the segments in the earthworm all work together for the whole organism. It is believed that segmentation evolved for many reasons, including a higher degree of motion. Taking the earthworm, for example: the segmentation of the muscular tissue allows the worm to move in an inching pattern. The circular muscles work to allow the segments to elongate one by one, and the longitudinal muscles then work to shorten the elongated segments. This pattern continues down the entirety of the worm, allowing it to inch along a surface. Each segment is allowed to work independently, but towards the movement of the whole worm.

- Heteronomous metamery is the condition where metameres have grouped together to perform similar tasks. The extreme example of this is the insect head (5 metameres), thorax (3 metameres), and abdomen (11 metameres, not all discernible in all insects). The process that results in the grouping of metameres is called "tagmatization", and each grouping is called a tagma (plural: tagmata). In organisms with highly derived tagmata, such as the insects, much of the metamerism with in a tagma may not be trivially distinguishable. It may have to be sought in structures that do not necessarily reflect the grouped metameric function (eg. the ladder nerve system or somites do not reflect the unitary structure of a thorax).

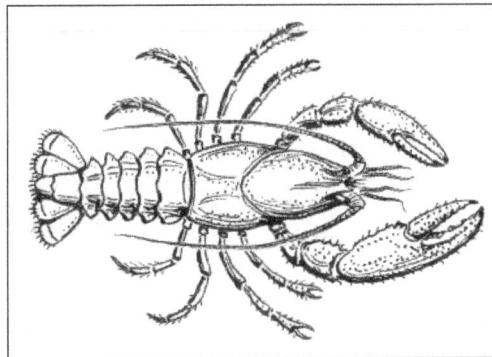

Segments of a crayfish exhibit metamerism.

In addition, an animal may be classified as "pseudometameric", meaning that it has clear internal metamerism but no corresponding external metamerism - as is seen, for example, in Monoplacophora.

Humans and other chordates are conspicuous examples of organisms that have metameres intimately grouped into tagmata. In the Chordata the metameres of each tagma are fused to such an extent that few repetitive features are directly visible. Intensive investigation is necessary to discern the metamerism in the tagmata of such organisms. Examples of detectable evidence of vestigially metameric structures include branchial arches and cranial nerves.

Some schemes regard the concept of metamerism as one of the four principles of construction of the human body, common to many animals, along with general bilateral symmetry (or zygomorphism), pachymerism (or tubulation), and stratification. More recent schemes also include three other concepts: segmentation (conceived as different from metamerism), polarity and endocrinosity.

In Plants

A metamer is one of several segments that share in the construction of a shoot, or into which a shoot may be conceptually (at least) resolved. In the metameristic model, a plant consists of a series of 'phytons' or phytomers, each consisting of an internode and its upper node with the attached leaf. As Asa Gray (1850) wrote:

The branch, or simple stem itself, is manifestly an assemblage of similar parts, placed one above another in a continuous series, developed one from another in successive generations. Each one of these joints of stem, bearing its leaf at the apex, is a plant element; or as we term it a phyton,—a potential plant, having all the organs of vegetation, namely, stem, leaf, and in its downward development even a root, or its equivalent. This view of the composition of the plant, though by no means a new one, has not been duly appreciated. I deem it essential to a correct philosophical understanding of the plant.

Some plants, particularly grasses, demonstrate a rather clear metameric construction, but many others either lack discrete modules or their presence is more arguable. Phyton theory has been criticized as an over-ingenious, academic conception which bears little relation to reality. Eames (1961) concluded that "concepts of the shoot as consisting of a series of structural units have been obscured by the dominance of the stem- and leaf-theory. Anatomical units like these do not exist: the shoot is the basic unit." Even so, others still consider comparative study along the length of the metameric organism to be a fundamental aspect of plant morphology.

Metameric conceptions generally segment the vegetative axis into repeating units along its length, but constructs based on other divisions are possible. The pipe model theory conceives of the plant (especially trees) as made up of unit pipes ('metamers'), each supporting a unit amount of photosynthetic tissue. Vertical metamers are also suggested in some desert shrubs in which the stem is modified into isolated strips of xylem, each having continuity from root to shoot. This may enable the plant to abscise a large part of its shoot system in response to drought, without damaging the remaining part.

In vascular plants, the shoot system differs fundamentally from the root system in that the former shows a metameric construction (repeated units of organs; stem, leaf, and inflorescence), while the latter does not. The plant embryo represents the first metamer of the shoot in spermatophytes or seed plants.

Plants (especially trees) are considered to have a 'modular construction,' a module being an axis in which the entire sequence of aerial differentiation is carried out from the initiation of the meristem to the onset of sexuality (e.g. flower or cone development) which completes its development. These modules are considered to be developmental units, not necessarily structural.

Metamorphosis

Metamorphosis is a biological process by which an animal physically develops after birth or hatching, involving a conspicuous and relatively abrupt change in the animal's body structure through cell growth and differentiation. Some insects, fishes, amphibians, mollusks, crustaceans, cnidarians, echinoderms, and tunicates undergo metamorphosis, which is often accompanied by a change of nutrition source or behavior. Animals can be divided into species that undergo complete metamorphosis ("holometaboly"), incomplete metamorphosis ("hemimetaboly"), or no metamorphosis ("ametaboly").

A dragonfly in its final moult, undergoing metamorphosis from its nymph form to an adult.

Scientific usage of the term is technically precise, and it is not applied to general aspects of cell growth, including rapid growth spurts. References to "metamorphosis" in mammals are imprecise and only colloquial, but historically idealist ideas of transformation and monadology, as in Goethe's Metamorphosis of Plants, have influenced the development of ideas of evolution.

Hormonal Control

Metamorphosis is iodothyronine-induced and an ancestral feature of all chordates. In insects growth and metamorphosis are controlled by hormones synthesized by endocrine glands near the front of the body (anterior). Neurosecretory cells in an insect's brain secrete a hormone, the prothoracicotropic hormone (PTTH) that activates prothoracic glands, which secrete a second hormone, usually ecdysone (an ecdysteroid), that induces ecdysis. PTTH also stimulates the corpora allata, a retrocerebral organ, to produce juvenile hormone, which prevents the development of adult characteristics during ecdysis. In holometabolous insects, molts between larval instars have a high level of juvenile hormone, the moult to the pupal stage has a low level of juvenile hormone, and the final, or imaginal, molt has no juvenile hormone present at all. Experiments on firebugs have shown how juvenile hormone can affect the number of nymph instar stages in hemimetabolous insects.

Insects

All three categories of metamorphosis can be found in the diversity of insects, including no metamorphosis ("ametaboly"), incomplete or partial metamorphosis ("hemimetaboly"), and complete metamorphosis ("holometaboly"). While ametabolous insects show very little difference between larval and adult forms (also known as "direct development"), both hemimetabolous and holometabolous insects have significant morphological and behavioral differences between larval and adult forms, the most significant being the inclusion, in holometabolus organisms, of a pupal or resting stage between the larval and adult forms.

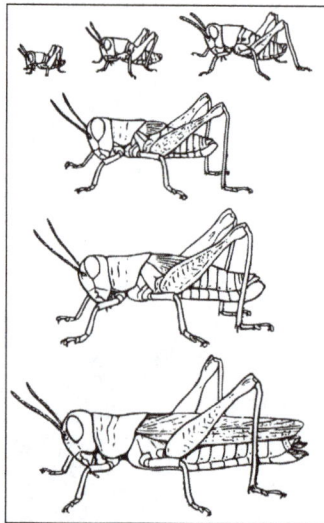

Incomplete metamorphosis in the grasshopper with different instar nymphs.

Development and Terminology

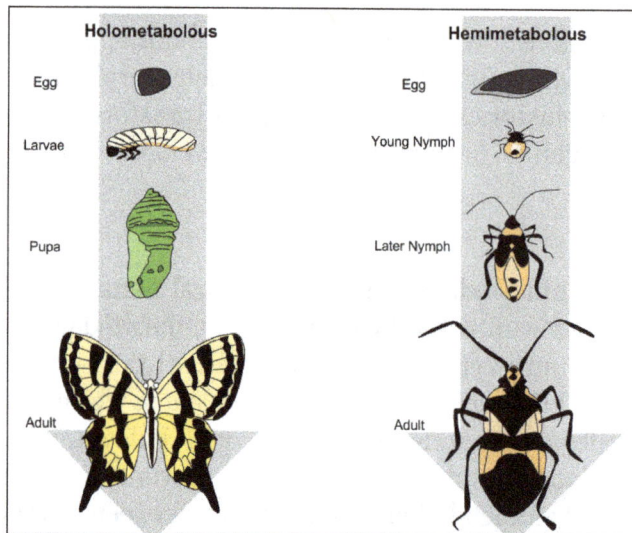

Two types of metamorphosis are shown. In a complete (holometabolous) metamorphosis the insect passes through four distinct phases, which produce an adult that does not resemble the larva. In an incomplete (hemimetabolous) metamorphosis an insect does not go through a full transformation, but instead transitions from a nymph to an adult by molting its exoskeleton as it grows.

In hemimetabolous insects, immature stages are called nymphs. Development proceeds in repeated stages of growth and ecdysis (moulting); these stages are called instars. The juvenile forms closely resemble adults, but are smaller and lack adult features such as wings and genitalia. The size and morphological differences between nymphs in different instars are small, often just differences in body proportions and the number of segments; in later instars, external wing buds form.

In holometabolous insects, immature stages are called larvae and differ markedly from adults. Insects which undergo holometabolism pass through a larval stage, then enter an inactive state called pupa (called a "chrysalis" in butterfly species), and finally emerge as adults.

Evolution

The earliest insect forms showed direct development (ametabolism), and the evolution of metamorphosis in insects is thought to have fuelled their dramatic radiation (1,2). Some early ametabolous "true insects" are still present today, such as bristletails and silverfish. Hemimetabolous insects include cockroaches, grasshoppers, dragonflies, and true bugs. Phylogenetically, all insects in the Pterygota undergo a marked change in form, texture and physical appearance from immature stage to adult. These insects either have hemimetabolous development, and undergo an incomplete or partial metamorphosis, or holometabolous development, which undergo a complete metamorphosis, including a pupal or resting stage between the larval and adult forms.

A number of hypotheses have been proposed to explain the evolution of holometaboly from hemimetaboly, mostly centering on whether or not the intermediate hemimetabolous forms are homologous to pupal form of holometabolous forms.

More recently, scientific attention has turned to characterizing the mechanistic basis of metamorphosis in terms of its hormonal control, by characterizing spatial and temporal patterns of hormone expression relative to metamorphosis in a wide range of insects.

Recent Research

According to research from 2008, adult *Manduca sexta* is able to retain behavior learned as a caterpillar. Another caterpillar, the ornate moth caterpillar, is able to carry toxins that it acquires from its diet through metamorphosis and into adulthood, where the toxins still serve for protection against predators.

Many observations have indicated that programmed cell death plays a considerable role during physiological processes of multicellular organisms, particularly during embryogenesis and metamorphosis.

The sequence of the metamorphosis of the butterfly:

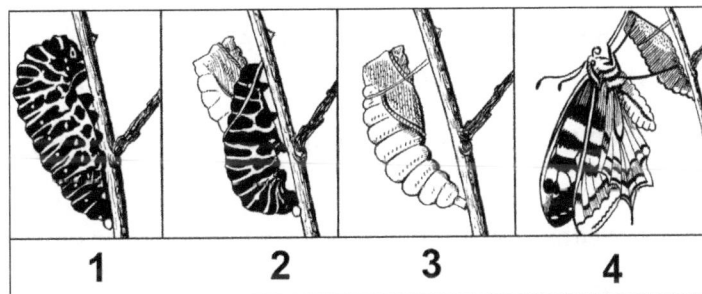

Metamorphosis of butterfly (PSF).

- The larva of a butterfly,
- The pupa is now spewing the thread to form cocoon,

- The coccoon is fully formed,

- Adult butterfly coming out of the cocoon.

Sequence illustrating complete metamorphosis in the cabbage white butterfly, Pieris rapae:

Larva

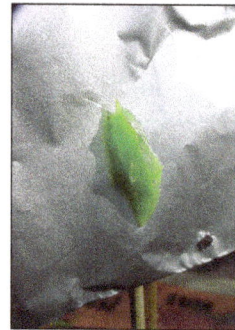
Pupa

Adult

Pupa ready to hatch

Chordata

Amphioxus

In cephalochordata, metamorphosis is iodothyronine-induced and it could be an ancestral feature of all chordates.

Fish

Some fish, both bony fish (Osteichthyes) and jawless fish (Agnatha), undergo metamorphosis. Fish metamorphosis is typically under strong control by the thyroid hormone.

Examples among the non-bony fish include the lamprey. Among the bony fish, mechanisms are varied.

The salmon is diadromous, meaning that it changes from a freshwater to a saltwater lifestyle.

Many species of flatfish begin their life bilaterally symmetrical, with an eye on either side of the body; but one eye moves to join the other side of the fish – which becomes the upper side – in the adult form.

The European eel has a number of metamorphoses, from the larval stage to the leptocephalus stage, then a quick metamorphosis to glass eel at the edge of the continental shelf (eight days for the Japanese eel), two months at the border of fresh and salt water where the glass eel undergoes a quick metamorphosis into elver, then a long stage of growth followed by a more gradual metamorphosis to the migrating phase. In the pre-adult freshwater stage, the eel also has phenotypic plasticity because fish-eating eels develop very wide mandibles, making the head look blunt. Leptocephali are common, occurring in all Elopomorpha (tarpon- and eel-like fish).

Most other bony fish undergo metamorphosis from embryo to larva (fry) and then to the juvenile stage during absorption of the yolk sac, because after that phase the individual needs to be able to feed for itself.

Amphibians

Just before metamorphosis, only 24 hours are needed to reach the stage in the next picture.

Almost functional common frog with some remains of the gill sac and a not fully developed jaw.

In typical amphibian development, eggs are laid in water and larvae are adapted to an aquatic lifestyle. Frogs, toads, and newts all hatch from the eggs as larvae with external gills but it will take some time for the amphibians to interact outside with pulmonary respiration. Afterwards, newt larvae start a predatory lifestyle, while tadpoles mostly scrape food off surfaces with their horny tooth ridges.

Metamorphosis in amphibians is regulated by thyroxin concentration in the blood, which stimulates metamorphosis, and prolactin, which counteracts its effect. Specific events are dependent on threshold values for different tissues. Because most embryonic development is outside the parental body, development is subject to many adaptations due to specific ecological circumstances. For this reason tadpoles can have horny ridges for teeth, whiskers, and fins. They also make use of the lateral line organ. After metamorphosis, these organs become redundant and will be resorbed by controlled cell death, called apoptosis. The amount of adaptation to specific ecological circumstances is remarkable, with many discoveries still being made.

Frogs and Toads

With frogs and toads, the external gills of the newly hatched tadpole are covered with a gill sac after a few days, and lungs are quickly formed. Front legs are formed under the gill sac, and hindlegs are visible a few days later. Following that there is usually a longer stage during which the tadpole lives off a vegetarian diet. Tadpoles use a relatively long, spiral-shaped gut to digest that diet.

Rapid changes in the body can then be observed as the lifestyle of the frog changes completely. The spiral-shaped mouth with horny tooth ridges is resorbed together with the spiral gut. The animal develops a big jaw, and its gills disappear along with its gill sac. Eyes and legs grow quickly, a tongue is formed, and all this is accompanied by associated changes in the neural networks (development of stereoscopic vision, loss of the lateral line system, etc.) All this can happen in about a day, so it is truly a metamorphosis. It is not until a few days later that the tail is reabsorbed, due to the higher thyroxin concentrations required for tail resorption.

Salamanders

The Salamander development is highly diverse; some species go through a dramatic reorganization when transitioning from aquatic larvae to terrestrial adults, while others, such as the Axolotl, display paedomorphosis and never develop into terrestrial adults. Within the genus *Ambystoma*, species have evolved to be paedomorphic several times, and paedomorphosis and complete development can both occur in some species.

Newts

The large external gills of the crested newt.

In newts, there is no true metamorphosis because newt larvae already feed as predators and continue doing so as adults. Newts' gills are never covered by a gill sac and will be resorbed only just before the animal leaves the water. Just as in tadpoles, their lungs are functional early, but newts use them less frequently than tadpoles. Newts often have an aquatic phase in spring and summer, and a land phase in winter. For adaptation to a water phase, prolactin is the required hormone, and for adaptation to the land phase, thyroxin. External gills do not return in subsequent aquatic phases because these are completely absorbed upon leaving the water for the first time.

Caecilians

Basal caecilians such as Ichthyophis go through a metamorphosis in which aquatic larva transition into fossorial adults, which involves a loss of the lateral line. More recently diverged caecilians (the Teresomata) do not undergo an ontogenetic niche shift of this sort and are in general fossorial throughout their lives. Thus, most caecilians do not undergo an anuran-like metamorphosis.

Ontogeny

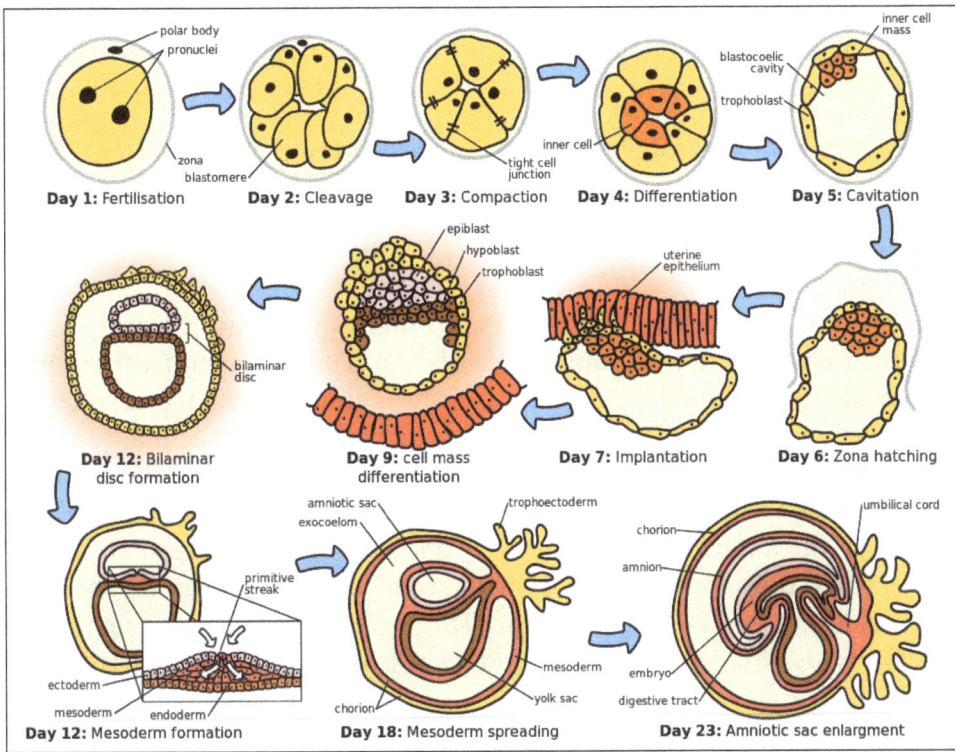

The initial stages of human embryogenesis.

Ontogeny is the origination and development of an organism (both physical and psychological, e.g., moral development), usually from the time of fertilization of the egg to the organism's mature form—although the term can be used to refer to the study of the entirety of an organism's lifespan.

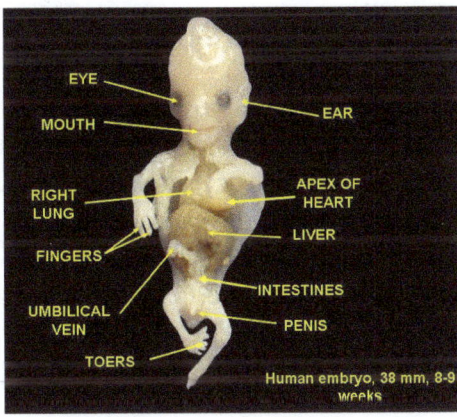

Parts of a human embryo.

Ontogeny is the developmental history of an organism within its own lifetime, as distinct from phylogeny, which refers to the evolutionary history of a species. In practice, writers on evolution often speak of species as "developing" traits or characteristics. This can be misleading. While developmental (i.e., ontogenetic) processes can influence subsequent evolutionary (e.g., phylogenetic) processes , individual organisms develop (ontogeny), while species evolve (phylogeny).

Ontogeny, embryology and developmental biology are closely related studies and those terms are sometimes used interchangeably. The term ontogeny has also been used in cell biology to describe the development of various cell types within an organism.

It is a useful field of study in many disciplines, including developmental biology, developmental psychology, developmental cognitive neuroscience, and developmental psychobiology.

It is also a concept used in anthropology as "the process through which each of us embodies the history of our own making".

Nature and Nurture

A seminal paper named ontogeny as one of the four primary questions of biology, along with Huxley's three others: causation, survival value and evolution. Tinbergen emphasized that the change of behavioral machinery during development was distinct from the change in behavior during development. "We can conclude that the thrush itself, i.e. its behavioral machinery, has changed only if the behavior change occurred while the environment was held constant. When we turn from description to causal analysis, and ask in what way the observed change in behavior machinery has been brought about, the natural first step is to try and distinguish between environmental influences and those within the animal. In ontogeny the conclusion that a certain change is internally controlled (is "innate") is reached by elimination. Tinbergen was concerned that the elimination of environmental factors is difficult to establish, and the use of the word "innate" is often misleading.

Ontogenetic Allometry

Most organisms undergo allometric changes in shape as they grow and mature, while others engage in metamorphosis. Even "reptiles" (non-avian sauropsids, e.g., crocodilians, turtles, snakes, lizards), in which the offspring are often viewed as miniature adults, show a variety of ontogenetic changes in morphology and physiology.

Anthropological Application

Comparing ourselves to others is something humans do all the time. "In doing so we are acknowledging not so much our sameness to others or our difference, but rather the commonality that resides in our difference. In other words, because each one of us is at once remarkably similar to, and remarkably different from, all other humans, it makes little sense to think of comparison in terms of a list of absolute similarities and a list of absolute differences. Rather, in respect of all other humans, we find similarities in the ways we are different from one another and differences in the ways we are the same. That we are able to do this is a function of the genuinely historical process that is human ontogeny".

Teratology

Teratology is the study of abnormalities of physiological development. It is often thought of as the study of human congenital abnormalities, but it is broader than that, taking into account other non-birth developmental stages, including puberty; and other organisms, including plants. The related term developmental toxicity includes all manifestations of abnormal development that are caused by environmental insult. These may include growth retardation, delayed mental development or other congenital disorders without any structural malformations. Teratogens are substances that may cause birth defects via a toxic effect on an embryo or fetus.

Mammalia

Teratogenesis

Along with this new awareness of the in utero vulnerability of the developing mammalian embryo came the development and refinement of The Six Principles of Teratology which are still applied today. These principles of teratology were put forth by Jim Wilson in 1959 and in his monograph Environment and Birth Defects. These principles guide the study and understanding of teratogenic agents and their effects on developing organisms:

- Susceptibility to teratogenesis depends on the genotype of the conceptus and the manner in which this interacts with adverse environmental factors.

- Susceptibility to teratogenesis varies with the developmental stage at the time of exposure to an adverse influence. There are critical periods of susceptibility to agents and organ systems affected by these agents.

- Teratogenic agents act in specific ways on developing cells and tissues to initiate sequences of abnormal developmental events.

- The access of adverse influences to developing tissues depends on the nature of the influence. Several factors affect the ability of a teratogen to contact a developing conceptus, such as the nature of the agent itself, route and degree of maternal exposure, rate of placental transfer and systemic absorption, and composition of the maternal and embryonic/ fetal genotypes.

- There are four manifestations of deviant development (Death, Malformation, Growth Retardation and Functional Defect).

- Manifestations of deviant development increase in frequency and degree as dosage increases from the No Observable Adverse Effect Level (NOAEL) to a dose producing 100% Lethality (LD100).

Studies designed to test the teratogenic potential of environmental agents use animal model systems (e.g., rat, mouse, rabbit, dog, and monkey). Early teratologists exposed pregnant animals to environmental agents and observed the fetuses for gross visceral and skeletal abnormalities. While this is still part of the teratological evaluation procedures today, the field of Teratology is moving to a more molecular level, seeking the mechanism(s) of action by which these agents act. Genetically modified mice are commonly used for this purpose. In addition, pregnancy registries are large, prospective studies that monitor exposures women receive during their pregnancies and record the outcome of their births. These studies provide information about possible risks of medications or other exposures in human pregnancies.

Understanding how a teratogen causes its effect is not only important in preventing congenital abnormalities but also has the potential for developing new therapeutic drugs safe for use with pregnant women.

Humans

In humans, congenital disorders resulted in about 510,000 deaths globally in 2010. About 3% of newborns have a "major physical anomaly", meaning a physical anomaly that has cosmetic or functional significance.

Vaccinating while Pregnant

In humans, vaccination has become readily available, and is important to the prevention of some diseases like polio, rubella, and smallpox, among others. There has been no association between congenital malformations and vaccination, as shown in Finland in which expecting mothers received the oral polio vaccine and saw no difference in infant outcomes than mothers who had not received the vaccine. However, it is still not recommended to vaccinate for polio while pregnant unless there is risk of infection. Another important implication of this includes the ability to get the influenza vaccine while pregnant. During the 1918 and 1957 influenza pandemics, mortality in pregnant women was 45%. However, even with prevention through vaccination, influenza vaccination in pregnant women remains low at 12%. Munoz et al. demonstrated that there was no adverse outcomes observed in the new infants or mothers.

Causes

Causes of teratogenesis can broadly be classified as:

- Toxic substances, such as, for humans, drugs in pregnancy and environmental toxins in pregnancy:

  - Potassium iodide is a possible teratogen. Potassium iodide in its raw form is a mild irritant and should be handled with gloves. Chronic overexposure can have adverse effects on the thyroid.

- Vertically transmitted infection.

- Lack of nutrients. For example, lack of folate acid in the nutrition in pregnancy for humans can result in spina bifida. Folic acid is a synthetic form of folate acid. Folic is added to processed food products, such as flour and breakfast cereals. High levels of un-metabolized folic acid have been associated with several health problems.

- Physical restraint. An example is Potter syndrome due to oligohydramnios in humans.

- Genetic disorders.

- Alcohol consumption during pregnancy.

Other Animals

Fossil Record

Evidence for congenital deformities found in the fossil record is studied by paleopathologists, specialists in ancient disease and injury. Fossils bearing evidence of congenital deformity are scientifically significant because they can help scientists infer the evolutionary history of life's developmental processes. For instance, because a *Tyrannosaurus rex* specimen has been discovered with a block vertebra, it means that vertebrae have been developing the same basic way since at least the most recent common ancestor of dinosaurs and mammals. Other notable fossil deformities include a hatchling specimen of the bird-like dinosaur, *Troodon*, the tip of whose jaw was twisted. Another notably deformed fossil was a specimen of the choristodere *Hyphalosaurus*, which had two heads-the oldest known example of polycephaly.

Plantae

In botany, teratology investigates the theoretical implications of abnormal specimens. For example, the discovery of abnormal flowers—for example, flowers with leaves instead of petals, or flowers with staminoid pistils—furnished important evidence for the "foliar theory", the theory that all flower parts are highly specialised leaves.

Types of Deformations in Vegetals

Plants can have mutations that leads to different types of deformations such as:

- Fasciation: Development of the apex (growing tip) in a flat plane perpendicular to the axis of elongation.

- Variegation: Degenerescence of genes, manifesting itself among other things by anomalous pigmentation.

- Virescence: Anomalous development of a green pigmentation in unexpected parts of the plant.

- Phyllody: Floral organs or fruits are transformed into leaves.

- Witch's broom: Unusually high multiplication of branches in the upper part of the plant, mainly tree.

- Pelory: Zygomorphic flower regress to their ancestral actinomorph symetry.

- Proliferation: Repetitive growth of an entire organ like a flower.

Galls are not part of the vegetal teratology as they are outgrowth due to external factors like insects bites or parasites.

Regeneration

Regeneration is the process by which some organisms replace or restore lost or amputated body parts. Organisms differ markedly in their ability to regenerate parts. Some grow a new structure on the stump of the old one. By such regeneration whole organisms may dramatically replace substantial portions of themselves when they have been cut in two, or may grow organs or appendages that have been lost. Not all living things regenerate parts in this manner, however. The stump of an amputated structure may simply heal over without replacement. This wound healing is itself a kind of regeneration at the tissue level of organization: a cut surface heals over, a bone fracture knits, and cells replace themselves as the need arises.

Regeneration, as one aspect of the general process of growth, is a primary attribute of all living systems. Without it there could be no life, for the very maintenance of an organism depends upon the incessant turnover by which all tissues and organs constantly renew themselves. In some cases rather substantial quantities of tissues are replaced from time to time, as in the successive production of follicles in the ovary or the molting and replacement of hairs and feathers. More commonly, the turnover is expressed at the cellular level. In mammalian skin the epidermal cells produced in the basal layer may take several weeks to reach the outer surface and be sloughed off. In the lining of the intestines, the life span of an individual epithelial cell may be only a few days.

The motile, hairlike cilia and flagella of single-celled organisms are capable of regenerating themselves within an hour or two after amputation. Even in nerve cells, which cannot divide, there is an endless flow of cytoplasm from the cell body out into the nerve fibres themselves. New molecules are continuously being generated and degraded with turnover times measured in minutes or hours in the case of some enzymes, or several weeks as in the case of muscle proteins. (Evidently, the only molecule exempt from this inexorable turnover is deoxyribonucleic acid [DNA] which ultimately governs all life processes.)

There is a close correlation between regeneration and generation. The methods by which organisms reproduce themselves have much in common with regenerative processes. Vegetative reproduction, which occurs commonly in plants and occasionally in lower animals, is a process by which whole new organisms may be produced from fractions of parent organisms; *e.g.*, when a new plant develops from a cut portion of another plant, or when certain worms reproduce by splitting in two, each half then growing what was left behind. More commonly, of course, reproduction is achieved sexually by the union of an egg and sperm. Here is a case in which an entire organism develops from a single cell, the fertilized egg, or zygote. This remarkable event, which occurs in all organisms that reproduce sexually, testifies to the universality of regenerative processes. During the course of evolution the regenerative potential has not changed, but only the levels of organization at which it is expressed. If regeneration is an adaptive trait, it would be expected to occur more commonly among organisms that appear to have the greatest need of such a capability, either because the hazard of injury is great or the benefit to be gained is great.

The actual distribution of regeneration among living things, however, seems at first glance to be a rather fortuitous one. It is difficult indeed to understand why some flatworms are able to regenerate heads and tails from any level of amputation, while other species can regenerate in only one direction or are unable to regenerate at all. Why do leeches fail to regenerate, while their close relatives, the earthworms, are so facile at replacing lost parts? Certain species of insects regularly grow back missing legs, but many others are totally lacking in this capacity. Virtually all modern bony fishes can regenerate amputated fins, but the cartilaginous fishes (including the sharks and rays) are unable to do so. Among the amphibians, salamanders regularly regenerate their legs, which are not very useful for movement in their aquatic environment, while frogs and toads, which are so much more dependent on their legs, are nevertheless unable to replace them. If natural selection operates on the principle of efficiency, then it is difficult to explain these many inconsistencies.

Some cases are so clearly adaptive that there have evolved not only mechanisms for regeneration, but mechanisms for self-amputation, as if to exploit the regenerative capability. The process of losing a body part spontaneously is called autotomy. The division of a protozoan into two cells and the splitting of a worm into two halves may be regarded as cases of autotomy. Some colonial marine animals called hydroids shed their upper portions periodically. Many insects and crustaceans will spontaneously drop a leg or claw if it is pinched or injured. Lizards are famous for their ability to release their tails. Even the shedding of antlers by deer may be classified as an example of autotomy. In all these cases autotomy occurs at a predetermined point of breakage. It would seem that wherever nature contrives to lose a part voluntarily, it provides the capacity for replacement.

Sometimes, when part of a given tissue or organ is removed, no attempt is made to regenerate the lost structures. Instead, that which remains behind grows larger. Like regeneration, this phenomenon—known as compensatory hypertrophy—can take place only if some portion of the original structure is left to react to the loss. If three-quarters of the human liver is removed, for example, the remaining fraction enlarges to a mass equivalent to the original organ. The missing lobes of the liver are not themselves replaced, but the residual ones grow as large as necessary in order to restore the original function of the organ. Other mammalian organs exhibit similar reactions. The kidney, pancreas, thyroid, adrenal glands, gonads, and lungs compensate in varying degrees for reductions in mass by enlargement of the remaining parts.

It is not invariably necessary for the regenerating tissue to be derived from a remnant of the original tissue. Through a process called metaplasia, one tissue can be converted to another. In the case of lens regeneration in certain amphibians, in response to the loss of the original lens from the eye, a new lens develops from the tissues at the edge of the iris on the upper margin of the pupil. These cells of the iris, which normally contain pigment granules, lose their colour, proliferate rapidly, and collect into a spherical mass which differentiates into a new lens.

Modes of Regeneration

Basic Patterns

Not all organisms regenerate in the same way. In plants and in coelenterates such as the hydra and jellyfishes, missing parts are replaced by reorganization of preexisting ones. The wound is healed, and the neighbouring tissues reorganize themselves into whatever parts may have been cut off. This process of reorganization, called morphallaxis, is the most efficient way for simple organisms to regenerate. Higher animals, with more complex bodies, regenerate parts differently, usually by the production of a specialized bud, or blastema, at the site of amputation. The blastema, made up of cells that look very much alike despite their often diverse origins, made its first appearance evolutionarily in flatworms and is encountered in the regenerative processes of all higher animals. It provides the tissue that will form the regenerated part.

Atypical Regeneration

Sometimes the part that grows back is not the same as that which was lost, and, occasionally, regeneration may be induced without having lost anything at all. It is not uncommon for a regenerated part to be incomplete. Earthworms, for example, usually regenerate only five segments in the anterior direction even if more than that number have been amputated. Many insects regenerate abnormally small legs from which some segments may be missing. Tadpole tails when amputated grow back to about only half their original length. These and other cases testify to the fact that a little regeneration is often good enough—that it is not necessary in every case to reproduce a flawless copy of the original.

Sometimes that which is regenerated is very different from the original. Among the arthropods there are cases in which the stump of an antenna grows a leg, while a cut eyestalk regenerates an antenna. More commonly, the regenerated part may be a reasonable facsimile of the original but will differ in details. A regenerated lizard tail contains an unsegmented cartilaginous tube instead of a series of vertebrae as did the original tail. The spinal cord lacks segmented ganglia, and the scales in the skin differ in character from the original ones. A regenerated tail, therefore, is easily distinguished from an original one yet appears sufficient to serve the purpose. Another interesting case is that of jaw regeneration in salamanders. If the lower jaw is amputated a new one will grow back, but it is often smaller than the original. It contains teeth and a mandible, but lacks a new tongue. Furthermore, the new mandible is a cartilaginous model of the original, and is not known to convert into bone.

Sometimes more of a part grows back than has been removed by amputation. A limb stump, for example, can occasionally give rise to hands with extra digits. Lobsters have been known to regenerate double structures, in which case the new parts are mirror images of each other.

The Regeneration Process

Origin of Regeneration Material

Following amputation, an appendage capable of regeneration develops a blastema from tissues in the stump just behind the level of amputation. These tissues undergo drastic changes. Their cells, once specialized as muscle, bone, or cartilage, lose the characteristics by which they are normally identified (dedifferentiation); they then begin to migrate toward, and accumulate beneath, the wound epidermis, forming a rounded bud (blastema) that bulges out from the stump. Cells nearest the tip of the bud continue to multiply, while those situated closest to the old tissues of the stump differentiate into muscle or cartilage, depending upon their location. Development continues until the final structures at the tip of the regenerated appendage are differentiated, and all the proliferating cells are used up in the process.

The blastema cells seem to differentiate into the same kind of cells they were before, or into closely related types. Cells may perhaps change their roles under certain conditions, but apparently rarely do so. If a limb blastema is transplanted to the back of the same animal, it may continue its development into a limb. Similarly, a tail blastema transplanted elsewhere on the body will become a tail. Thus, the cells of a blastema seem to bear the indelible stamp of the appendage from which they were produced and into which they are destined to develop. If a tail blastema is transplanted to the stump of a limb, however, the structure that regenerates will be a composite of the two appendages.

Polarity and Gradient Theory

Each living thing exhibits polarity, one example of which is the differentiation of an organism into a head, or forward part, and a tail, or hind part. Regenerating parts are no exception; they exhibit polarity by always growing in a distal direction (away from the main part of the body). Among the lower invertebrates, however, the distinction between proximal (near, or toward the body) and distal is not always clear cut. It is not difficult, for example, to reverse the polarity of "stems" in colonial hydroids. Normally a piece of the stem will grow a head end, or hydranth, at its free, or distal, end; if that is tied off, however, it regenerates a hydranth at the end that was originally proximal. The polarity in this system is apparently determined by an activity gradient in such a way that a hydranth regenerates wherever the metabolic rate is highest. Once a hydranth has begun to develop, it inhibits the production of others proximal to it by the diffusion of an inhibitory substance downward along the stem.

When planarian flatworms are cut in half, each piece grows back the end that is missing. Cells in essentially identical regions of the body where the cut was made form blastemas, which, in one case gives rise to a head and in the other becomes a tail. What each blastema regenerates depends entirely on whether it is on a front piece or a hind piece of flatworm: the real difference between the two pieces may be established by metabolic differentials. If a transverse piece of a flatworm is cut very thin—too narrow for an effective metabolic gradient to be set up—it may regenerate two heads, one at either end. If the metabolic activity at the anterior end of a flatworm is artificially reduced by exposure to certain drugs, then the former posterior end of the worm may develop a head.

Appendage regeneration poses a different problem from that of whole organisms. The fin of a fish and the limb of a salamander have proximal and distal ends. By various manipulations, it is possible to make them regenerate in a proximal direction, however. If a square hole is cut in the fin of a fish, regeneration takes place as expected from the inner margin, but may also occur from the distal edge. In the latter case, the regenerating fin is actually a distal structure except that it happens to be growing in a proximal direction.

Amphibian limbs react in a similar manner. It is possible to graft the hand of a newt to the nearby body wall, and once a sufficient blood flow has been established, to sever the arm between the shoulder and elbow. This creates two stumps, a short one consisting of part of the upper arm, and a longer one made up of the rest of the arm protruding in the wrong direction from the side of the animal. Both stumps regenerate the same thing, namely, everything normally lying distal to the level of amputation, regardless of which way the stump was facing. The reversed arm therefore regenerates a mirror image of itself.

Clearly, when a structure regenerates it can only produce parts that normally lie distal to the level of amputation. The participating cells contain information needed to develop everything "downstream," but can never become more proximal structures. Regeneration, like embryonic development, occurs in a definite sequence.

Regulation of Regeneration

There are certain prerequisites without which regeneration cannot occur. First and foremost, there must be a wound, although the original appendage need not have been lost in the process. Second, there must be a source of blastema cells derived from remnants of the original structure or an associated one. Finally, regeneration must be stimulated by some external force. The stimuli often involve the nervous system. An adequate nerve supply is required for the regeneration of fish fins, taste barbels, and amphibian limbs. In the case of many tail regenerations, the spinal cord provides the necessary stimulus. Lens regeneration in salamander eyes depends upon the presence of a retina. Arthropod appendages regenerate in the presence of molting hormones. Protozoan regeneration requires the presence of a nucleus. In case after case, regeneration depends on more than a healed wound and a source of blastema cells. It is often triggered by some physiological stimulus originating elsewhere in the body, a stimulus invariably associated with the very function of the structure to be regenerated. The conclusion is inescapable that regeneration is primarily the recovery of deficient functions rather than simply the replacement of lost structures.

The imperative of need is of further importance in suppressing excess regeneration. To be able to regenerate is to run the risk of regenerating too much or too often. If regeneration did not depend upon a physiological stimulus, such as those mediated by nerves or hormones, there would be no reason why simple wounds should not sprout whole new appendages.

It is not known why regeneration fails to occur in many cases, as in the legs of frogs or the limbs and tails of mammals. The nerve supply might be inadequate, for when the number of nerves is artificially increased, regeneration is sometimes induced. This cannot be the whole answer, however, because not all appendages depend on nerves for their regeneration; newt jaws, salamander gills, and deer antlers do not require nerves to regenerate.

Possibly the failure to regenerate relates to the ways in which wounds heal. In higher vertebrates there is a tendency to form thick scar tissue in healing wounds, which may act as a barrier between the epidermis and the underlying tissues of the stump. In the absence of direct contact between these two tissues, the stump may not be able to give rise to the blastema cells required for regeneration.

The Range of Regenerative Capability

Virtually no group of organisms lacks the ability to regenerate something. This process, however, is developed to a remarkable degree in lower organisms, such as protists and plants, and even in many invertebrate animals such as earthworms and starfishes. Regeneration is much more restricted in higher organisms such as mammals, in which it is probably incompatible with the evolution of other body features of greater survival value to these complex animals.

Protists and Plants

Algae

One of the most outstanding feats of regeneration occurs in the single-celled green alga Acetabularia. This plant-like protist of shallow tropical water consists of a group of short rootlike appendages; a long thin "stem," up to several centimetres in length; and an umbrella-like cap at the top. The entire organism is one cell, with its single nucleus situated at the base in one of the "roots." If the cap is cut off, a new one regenerates from the healed over stump of the amputated stem. The nucleus is necessary for this kind of regeneration, presumably because it provides the information needed to direct the development of the new cap. Once this information has been produced by the nucleus, however, the nucleus can be removed and regeneration continues unabated.

If the nucleus from one species of Acetabularia is added to a cell-body of another species, and the cap of the recipient cell is amputated, the new cap that regenerates will be a hybrid because each nucleus exerts its own morphogenetic influences. On the other hand, if the nucleus from one species is substituted for that in another, regeneration reflects the properties of the new nucleus.

Protozoans

Most single-celled, animal-like protists regenerate very well. If part of the cell fluid, or cytoplasm, is removed from Amoeba, it is readily replaced. A similar process occurs in other protozoans, such as flagellates and ciliates. In each case, however, regeneration occurs only from that fragment of the cell containing the nucleus. Amputated parts that lack a nucleus cannot survive. In some ciliates, such as Blepharisma or Stentor, the nucleus may be elongated or shaped like a string of beads. If either of these organisms is cut in two so that each fragment retains part of the elongated nucleus, each half proceeds to grow back what it lacks, giving rise to a complete organism in less than six hours. The way in which such a bisected protozoan regenerates is almost identical with the way it reproduces by ordinary division. Even a very tiny fragment of the whole organism can regenerate itself, provided it contains some nuclear material to determine what is supposed to be regenerated.

Green Plants

The mechanisms by which vascular plants grow have much in common with regeneration. Their roots and shoots elongate by virtue of the cells in their meristems, the conical growth buds at the tip of each branch. These meristems are capable of indefinite growth, especially in perennial plants. If they are amputated they are not replaced, but other meristems along the stem, normally held in abeyance, begin to sprout into new branches that more than compensate for the loss of the original one. Such a process is called restitution.

Plants are also capable of producing callus tissue wherever they may be injured. This callus is proliferated from cambial cells, which lie beneath the surface of branches and are responsible for their increase in width. When a callus forms, some of its cells may organize into growing points, some of which in turn give rise to roots while others produce stems and leaves.

Invertebrates

Coelenterates

The vast majority of research on coelenterates has been focussed on hydras and some of the colonial hydroids. If a hydra is cut in half, the head end reconstitutes a new foot, while the basal portion regenerates a new hydranth with mouth and tentacles. This seemingly straightforward process is deceptively simple. From tiny fragments of the organism whole animals can be reconstituted. Even if a hydra is minced and the pieces scrambled, the fragments grow together and reorganize themselves into a complete whole. The indestructibility of the hydra may well be attributed to the fact that even the intact animal is constantly regenerating itself. Just below the mouth is a growth zone from which cells migrate into the tentacles and to the foot where they eventually die. Hence, the hydra is in a ceaseless state of turnover, with the loss of cells at the foot and at the tips of the tentacles being balanced by the production of new ones in the growth zone. If such an animal is X-rayed, the proliferation of new cells is inhibited and the hydra gradually shrinks and eventually dies owing to the inexorable demise of cells and the inability to replace them.

In colonial hydroids, such as Tubularia, there is a series of branching stems, each of which bears a hydranth on its end. If these hydranths are amputated they grow back within a few days. In fact, the organism normally sheds its hydranths from time to time and regenerates new ones naturally.

Flatworms

Planarian flatworms are well-known for their ability to regenerate heads and tails from cut ends. In the case of head regeneration, some blastema cells become brain tissues, others develop into the eyes, and still others differentiate as muscle or intestine. In a week or so, the new head functions almost as well as the original.

The blastema that normally gives rise to a single head is, under certain circumstances, even capable of becoming two heads if the stump of a decapitated flatworm is divided in two by a longitudinal cut. Each of the two halves then gives rise to a complete head. Thus, each blastema develops into an entire structure regardless of its size or position in relation to the rest of the animal.

In the case of flatworms there is still considerable disagreement concerning the origins of the blastema. Some investigators contend that it is derived from neoblasts, undifferentiated reserve cells scattered throughout the body. Others claim that there are no such reserve cells and that the blastema develops from formerly specialized cells near the wound that dedifferentiate to give rise to the blastema cells. Whatever their source, the cells of the blastema are capable of becoming many different things depending upon their location.

Regeneration in flatworms occurs in a stepwise fashion. The first tissue to differentiate is the brain, which induces the development of eyes. Once the head has formed, it in turn stimulates the production of the pharynx. The latter then induces the development of reproductive organs farther back. Thus, each part is necessary for the successful development of those to come after it; conversely, each part inhibits the production of more of itself. If decapitated flatworms are exposed to extracts of heads, the regeneration of their own heads is prevented. Such a complex interplay of stimulators and inhibitors is responsible for the successful regeneration of an integrated morphological structure.

Annelids

The segmented worms exhibit variable degrees of regeneration. The leeches, as already noted, are wholly lacking in the ability to replace lost segments, whereas the earthworms and various marine annelids (polychaetes) can often regenerate forward and backward. The expression of such regenerative capacities depends very much on the level of amputation. Anteriorly directed regeneration usually occurs best from cuts made through the front end of the worm, with little or no growth taking place from progressively more posterior bisections. Posteriorly directed regeneration is generally more common and extensive. Some species of worms replace the same number of segments as were lost. Hypomeric regeneration, in which fewer segments are produced than were removed, is more common, however.

Anterior regeneration depends upon the presence of the central nerve cord. If this is cut or deflected from the wound surface, little or no forward regeneration may take place. Posterior regeneration requires the presence of the intestine, removal of which precludes the formation of hind

segments. Thus, it would seem that no head will regenerate without a central nervous system, nor a tail without an opening.

Arthropods

Many insects and crustaceans regenerate legs, claws, or antennas with apparent ease. When insect legs regenerate, the new growth is not visible externally because it develops within the next proximal segment in the stump. Not until the following molt is it released from its confinement to unfold as a fully developed leg only slightly smaller than the original. In the case of crabs, regenerating legs bulge outward from the amputation stump. They are curled up within a cuticular sheath, not to be extended until the sheath is molted. Lobsters and crayfish regenerate claws and legs in a straightforward manner as direct outgrowths from the stumps. As in other crustaceans, however, these regenerates lie immobile within an enveloping cuticle and do not become functional until their sheath is shed at the next molt.

In all arthropods regeneration is associated with molting, and therefore takes place only during larval or young stages. Most insects do not initiate leg regeneration unless there remains ample time prior to the next scheduled molt for the new leg to complete its development. If amputation is performed too late in the intermolt period, the onset of regeneration is delayed until after shedding; the regenerate then does not appear until the second molt. Metamorphosis into the adult stage marks the end of molting in insects, and adults accordingly do not regenerate amputated appendages.

Crustaceans often tend to molt and grow throughout life. They therefore never lose the ability to grow back missing appendages. When a leg is lost, a new outgrowth appears even if the animal is not destined to molt for many months. Following a period of basal growth, during which a diminutive limb is produced, the regenerated part eventually ceases to elongate. Not until a few weeks before the next molt does it resume growth and complete its development, triggered by the hormones that induce molting.

Vertebrates

Fishes

Many different parts of the fish's body will grow back. Plucked scales are promptly replaced by new ones, and amputated gill filaments can regenerate easily. The "whiskers," or taste barbels, of the catfish grow back as perfect replicas of the originals. The most conspicuous regenerating structures in fishes, however, are the fins. When any of these are amputated, new fins grow out from the stumps and soon restore everything that was missing. Even the coloured stripes or spots that adorn some fins are reconstituted by new pigment cells that repopulate the regenerated part. Fin regeneration depends on an adequate nerve supply. If the nerves are cut leading into the fin, regeneration of neither the amputated fin nor excised pieces of the bony fin rays can take place.

Amphibians

Salamanders are remarkable for their ability to regenerate limbs. Larval frogs, or tadpoles, also possess this ability, but usually lose it when they become frogs. It is not known why frog legs do not regenerate, and under appropriate stimuli they can be induced to do so.

Tadpoles and salamanders can replace amputated tails. Tadpole tails have a stiff rod called the notochord for support, whereas salamanders possess a backbone, composed of vertebrae. Both tails contain a spinal cord. When the salamander regenerates its tail, the spinal cord grows back and segmental nerve-cell clusters (ganglia) differentiate. Tadpoles also regenerate their spinal cords, but not the associated ganglia. If the spinal cord is removed or destroyed in the salamander, no tail regeneration occurs; if it is removed from the tadpole tail, however, regeneration can proceed without it.

Reptiles

Lizards also regenerate their tails, especially in those species that have evolved a mechanism for breaking off the original tail when it is grasped by an enemy. When the lizard tail regenerates, however, it does not replace the segmented vertebrae. Instead, there develops a long tapering cartilaginous tube within which the spinal cord is located and outside of which are segmented muscles. The spinal cord of the lizard tail is necessary for regeneration, but the regenerated tail does not reproduce the ganglia that are normally associated with it. Occasionally, a side tail may be produced if the original tail is broken but not lost.

Birds

Regeneration of amputated appendages in birds is not known to occur; however, they do replace their feathers as a matter of course. While most species shed and regenerate feathers one at a time so as not to be grounded, flightless birds, such as penguins, may molt them all at once. Male puffins cast off their colorful beaks after the mating season, but grow new ones the following year. In like manner, the dorsal keel on the upper beaks of male pelicans is shed and replaced annually.

Mammals

Although mammals are incapable of regenerating limbs and tails, there are a few exceptional cases in which lost tissues are in fact regenerated. Not the least of these cases is the annual replacement of antlers in deer. These remarkable structures, which normally grow on the heads of male deer, consist of an inner core of bone enveloped by a layer of skin and nourished by a copious blood supply. During the growing season the antlers elongate by the proliferation of tissues at their growing tips. The rate of growth in some of the larger species may surpass one centimetre (0.39 inch) per day; the maximum rate of growth recorded for the elk is 2.75 centimetres (1.05 inches) per day. When the antlers have reached their full extent, the blood supply is constricted, and the skin, or velvet, peels off, thus revealing the hard, dead, bony antlers produced by the male deer in time for the autumn mating season. The regeneration of elk antlers spans about seven months. The following spring, the old antlers are shed and new ones grow to replace them.

Still another example of mammalian regeneration occurs in the case of the rabbit's ear. When a hole is punched through the external ear of the rabbit, tissue grows in from around the edges until the original opening is reduced or obliterated altogether. This regeneration is achieved by the production of new skin and cartilage from the margins of the original hole. A similar phenomenon occurs in the case of the bat's wing membrane.

References

- Ji Q.; Wu X.-C.; Cheng Y.-N. (2010). "Cretaceous choristoderan reptiles gave birth to live young". Naturwissen-schaften. 97 (4): 423–428. Doi:10.1007/s00114-010-0654-2. PMID 20179895

- E., Charles (2012-04-20). "Human Embryogenesis". Embryogenesis. Intech. Doi:10.5772/36871. ISBN 9789535104-667. Archived from the original on 2018-05-02

- Aging-life-process, science: britannica.com, Retrieved 14 July, 2019

- Irvine KD, Rauskolb C (2001). "Boundaries in development: formation and function". Annu Rev Cell Dev Biol. 17: 189–214. Doi:10.1146/annurev.cellbio.17.1.189. PMID 11687488

- Martin AC, Wieschaus EF (2010). "Tensions divide". Nat Cell Biol. 12 (1): 5–7. Doi:10.1038/ncb0110-5. PMID 20027198

- Lois N. Magner (2005). History of the Life Science s. New York: Marcel Dekker. P. 166. ISBN 9780824743604

- Laudet, Vincent (September 27, 2011). "The Origins and Evolution of Vertebrate Metamorphosis". Current Biology. 21 (18): R726–R737. Doi:10.1016/j.cub.2011.07.030. PMID 21959163

- Regeneration-biology, science: britannica.com, Retrieved 31 March, 2019

Development Processes in Animals

3

- **Mating**
- **Fertilization**
- **Cleavage**
- **Blastulation**
- **Implantation**
- **Embryonic Disc**
- **Gastrulation**
- **Neurulation**

The process which leads to the formation of a new animal and which begins with the derivation of cells from one or more parent individuals is known as animal development. It involves the processes such as mating, fertilization, blastulation, implantation, etc. The topics elaborated in this chapter will help in gaining a better perspective about these processes related to animal development.

Animal development is the processes that lead eventually to the formation of a new animal starting from cells derived from one or more parent individuals. Development thus occurs following the process by which a new generation of organisms is produced by the parent generation.

Features

Reproduction and Development

In multicellular animals (Metazoa), reproduction takes one of two essentially different forms: sexual and asexual. In asexual reproduction the new individual is derived from a blastema, a group of cells from the parent body, sometimes, as in Hydra and other coelenterates, in the form of a "bud" on the body surface. In sponges and bryozoans, the cell groups from which new individuals develop are formed internally and may be surrounded by protective shells; these bodies, which may serve as resistant forms capable of withstanding unfavourable environmental conditions, are released after the death of the parent. In certain animals the parent may split in half, as in some worms, in which an individual worm breaks into two fairly equal parts (except that the anterior half receives the mouth, "brain," and sense organs if they are present).

Obviously, in such a case it is impossible to say which of the two resulting individuals is the parent and which the offspring. Some brittle stars (starfish relatives) may reproduce by breaking across the middle of the body disk, with each of the halves subsequently growing its missing half and the corresponding arms.

A common feature of all forms of asexual reproduction is that the cells—always a substantial number of cells, never only one cell—taking part in the formation of the new individual are not essentially different from other body, or somatic, cells. The number of chromosomes (bodies carrying the hereditary material) in the cells participating in the formation of a blastema is the same as in the other somatic cells of the parent, constituting a normal, double, or diploid ($2n$), set.

In sexual reproduction, a new individual is produced not by somatic cells of the parent but by sex cells, or gametes, which differ essentially from somatic cells in having undergone meiosis, a process in which the number of chromosomes is reduced to one-half of the diploid ($2n$) number found in somatic cells; cells containing one set of chromosomes are said to be haploid (n). The resulting sex cells thus receive only half the number of chromosomes present in the somatic cell. Furthermore, the sex cells are generally capable of developing into a new individual only after two have united in a process called fertilization.

Each type of reproduction—asexual and sexual—has advantages for the species. Asexual reproduction is, at least in some cases, the faster process, leading most rapidly to the development of large numbers of individuals. Males and females are independently capable of producing offspring. The large size of the original mass of living matter and its high degree of organization—the new individual inherits parts of the body of the parent: a part of the alimentary canal, for instance—make subsequent development more simple, and the attainment of a stage capable of self-support easier. New individuals produced by asexual reproduction have the same genetic constitution (genotype) as their parent and constitute what is called a clone. Though asexual reproduction is advantageous in that, if the parent animal is well adapted to its environment and the latter is stable, then all offspring will benefit, it is disadvantageous in that the fixed genotype not only makes any change in offspring impossible, should the environment change, but also prevents the acquisition of new characteristics, as part of an evolutionary process. Sexual reproduction, on the other hand, provides possibilities for variation among offspring and thus assists evolution by allowing new pairs of genes to combine in offspring. Since all body cells are derived from the fertilized egg cell, a mutation, or change, occurring in the sex cells of the parents immediately provides a new genotype in each cell of the offspring. In the course of evolution, sexual reproduction has been selected for, and established in, all main lines of organisms; asexual reproduction is found only in special cases and restricted groups of organisms.

Preparatory Events

In the case of multicellular animals we find there are two kinds of sex cells: the female sex cell (ovum, or egg), derived from an oocyte (immature egg), and the male sex cell (spermatozoon or sperm), derived from a spermatocyte. Eggs are produced in ovaries; sperm, in testes. Both the egg and the sperm contribute to the development of the new individual; each providing one set of genes, thereby restoring the diploid number of chromosomes in the fertilized egg. The sperm possesses a whiplike tail (flagellum) that enables it to swim to the egg to fertilize it. In most cases the egg, a stationary, spherical cell, provides the potential offspring with a store of food materials, or yolk, for its early development. The term yolk does not refer to any particular substance but in fact includes proteins, phosphoproteins, lipids, cholesterol, and fats, all of which substances occur in various proportions in the eggs of different animals. In addition to yolk, eggs accumulate other components and acquire the structure necessary for the development of the new individual. In particular the egg acquires polarity—that is, the two ends, or poles, of the egg become distinctive from each other. At one pole, known as the animal pole, the cytoplasm appears to be more active and contains the nucleus (meiotic divisions occur in this region); at the other, called the vegetal pole, the cytoplasm is less active and contains most of the yolk. The general organization of the future animal is closely related to the polarity of the egg.

When the amount of food reserve is comparatively small, as it is in many marine invertebrates and mammals (in the latter the embryo is nourished by materials in the mother's blood), the egg may be barely visible to the unaided eye. The egg of the sea urchin is about 75 microns (0.003 inch) in diameter; that of a human being is slightly more than 0.1 millimetre. Eggs are classified according to the amount of yolk present. An egg with a small quantity of evenly distributed yolk is called an oligolecithal egg. One with more yolk that is unevenly distributed (*i.e.*, concentrated towards the vegetal pole) is telolecithal; and one with still greater amounts of yolk in granules or in a compact mass is megalecithal.

The egg is surrounded by protective membranes, which may be soft and jellylike or hard and calcified, like shells. Egg membranes are produced while the egg is either in the ovary or being carried away from the ovary in a tube called an oviduct. The eggs of many animals have both kinds of membranes. In insects, a hard shell (chorion) forms around the eggs in the ovaries. In frogs, a very thin vitelline membrane forms around the eggs in the ovary; subsequently a layer of jelly is deposited around the eggs while they pass through the oviducts. In birds, a very thin vitelline membrane is produced around the egg in the ovary; then several layers of secondary membranes are formed in the oviduct before the egg is laid. The outermost of these secondary membranes is the calcareous shell. In mammals the egg is surrounded by the so-called pellucid zone, which is equivalent to the vitelline membrane of other animals; follicle cells form an area called the corona radiata around this zone.

After fertilization the egg, now called a zygote, is endowed with genes from two parents and has begun actual development. (Activation of the egg may be brought about by an agent other than sperm in certain animals, but such cases of parthenogenesis are exceptional.

After fertilization, the zygote undergoes a series of transformations that bring it closer to the essential organization of the parents. These transformations, initiated at a physiological, perhaps even at a molecular, level, eventually result in the appearance of certain structures. The whole process is called morphogenesis. The process of development is more easily understood if, at every step, the changes necessary to bring the system nearer the goal are considered. Depending on the achievements necessary at any step, development can be subdivided into a number of discrete phases, the first of which, cleavage, immediately follows fertilization.

Early Development

Amphioxus, Echinoderms and Amphibians

In the course of evolution, certain animal groups have modified this critical stage of embryonic development, and these modifications have undoubtedly contributed to the successful continuation of species. In the primitive fishlike chordate amphioxus, for example, the invaginating blastoderm eventually comes into close contact with the inner surface of the ectoderm, thus practically squeezing the blastocoel out of existence or at least reducing it to a narrow crevice between the ectoderm and the endomesoderm. In echinoderms, on the other hand, a smaller portion of the blastoderm invaginates, and the blastocoel remains as a spacious internal cavity between the ectoderm and the endomesoderm. It persists as the primary body cavity and is the only body cavity (apart from the cavity of the alimentary canal) in such invertebrates as nematodes and rotifers.

In the double-walled-cup stage, the two internal germinal layers—endoderm and mesoderm—may not yet be distinct. Their separation may occur later, in the second phase of gastrulation, by one of two methods. One is the development of outpocketings from the wall of the archenteron. In starfishes and other echinoderms, the deep part of the endomesodermal invagination forms two thin-walled sacs, one on each side of the gastrula. These are the rudiments of the mesoderm; the remaining part of the archenteron becomes the endoderm and produces the lining of the gut. The cavities within the mesodermal sacs expand to become the coelom, the secondary body cavity of the animal. A somewhat similar process of mesoderm and coelom development occurs in amphioxus among the chordates, except that a series of mesodermal sacs forms on either side of the embryo, foreshadowing the segmented (metameric) structure common to chordates. Only the most anterior pairs of the mesodermal sacs actually contain a cavity at the time of their formation; the more posterior ones are solid masses of cells separating from the archenteric wall and from one another and developing coelomic cavities later.

A second method of mesoderm formation is by the splitting off of mesodermal cells from the original common mass of endomesoderm. This may take the form of single cells detaching themselves from the archenteron or of whole sheets of cells splitting off from the endoderm. An example of the latter type is seen in the gastrulation of amphibians. The development of specific regions of the early amphibian embryo—by the use of natural pigmentation or artificially introduced dyes—can be followed and their location in the adult recorded in diagrams called fate maps. The fate map of a frog blastula just prior to gastrulation demonstrates that the materials for the various organs of the embryo are not yet in the position corresponding to that in which the organs will lie in a fully developed animal. The endodermal material for the foregut, for example, lies not far from the vegetal pole; the ectodermal component of the mouth region (stomodeum) is situated close to the animal pole. Extensive rearrangement of the embryo is necessary to bring all the parts into their correct relationships.

Because of the large amount of yolk and resulting uneven cleavage, gastrulation in amphibians cannot proceed by a simple infolding of the vegetal hemisphere. A certain amount of invagination does take place, assisted by an active spreading of the animal hemisphere of the embryo; as a result, the ectoderm covers the endodermal and mesodermal areas. The spreading is sometimes described as an "overgrowth"—an inappropriate term, since no growth or increase of mass is involved. The future ectoderm simply thins out, expands, and covers a greater surface of the embryo in a movement known as epiboly.

Gastrulation in amphibians, in lungfishes, and in the cyclostomes (hagfishes and lampreys) begins with the formation of a pit on what will become the back (dorsal) side of the embryo. The pit represents the active shifting inward of the cells of the blastoderm. As these cells undergo a change in shape, there occurs also a contraction at the external surface, with adjacent cells being drawn toward the centre of the contraction even before an actual depression is formed. The cells most concerned in this process will become part of the future foregut. Further movement of the cells inward results in the formation of a distinct pit, which rapidly develops into a pocket-like archenteron with its opening, the blastopore. Once the archenteron is formed, more and more of the exterior cells roll over the edge of the blastopore and disappear into the interior. In the course of gastrulation the shape of the blastopore changes from a simple pit to a transverse slit and finally into a groove encircling the yolky material at the vegetal pole. As a result of epiboly of the animal hemisphere, the upper edge of the groove is gradually pushed down until the yolky cells of the vegetal pole are covered completely. The edges of the blastopore then converge toward the vegetal pole, the slit between them being eventually reduced to a narrow canal, which lies at the posterior end of the embryo and, in some species, becomes the anal opening. (In other cases the canal closes, and a new anal opening breaks through nearby, slightly more ventrally.)

The cavity of the archenteron increases as more material from the outside is transferred inward, and the blastocoel becomes almost completely obliterated. Both mesoderm and endoderm are shifted into the interior, and only the ectoderm remains on the embryo surface. The mesoderm splits from the endoderm: the endoderm lines the archenteric cavity (and eventually becomes the lining of the alimentary canal), as the mesoderm surrounds the endoderm to form the chordamesodermal mantle. By the time the blastopore closes, the three germ layers are in their correct spatial relationship to each other.

Reptiles, Birds and Mammals

Although amphibian gastrulation is considerably modified in comparison with that in animals with oligolecithal eggs (*e.g.,* amphioxus and starfishes), an archenteron forms by a process of invagination. Such is not the case, however, in the higher vertebrates that possess eggs with enormous amounts of yolk, as do the reptiles, birds, and egg-laying mammals. Cleavage in these animals is partial (meroblastic), and, at its conclusion, the embryo consists of a disk-shaped group of cells lying on top of a mass of yolk. This cell group often splits into an upper layer, the epiblast, and a lower layer, the hypoblast. These layers do not represent ectoderm and endoderm, respectively, since almost all the cells that form the embryo are contained in the epiblast. Future mesodermal and endodermal cells sink down into the interior, leaving only the ectodermal material at the surface. In reptiles, egg-laying mammals, and some birds, a pocket-like depression occurs in the epiblast but encompasses only chordamesoderm or even only the notochord. Individual cells of the remainder of the mesoderm and endoderm migrate into the interior and there arrange themselves into a sheet of chordamesoderm and of endoderm, the latter of which mingles with cells of the hypoblast if such a layer is present. The migration of the cells destined to form mesoderm and endoderm does not take place over the whole surface of the disk-shaped embryo but is restricted to a specific area along the midline. This area is more or less oval in reptiles and lower mammals; distinctly elongated in higher mammals and birds, it is called the primitive streak, a thickened and slightly depressed part of the epiblast that is thickest at the anterior end, called the Hensen's node.

In animals having discoidal cleavage, the three germinal layers at the end of gastrulation are stacked flat; ectoderm on top, mesoderm in the middle, and endoderm at the bottom. The embryo is produced from the flattened layers by a process of folding to form a system of concentric tubes. The edges of the germ layers, which are not involved in the folding process, remain attached to the yolk and become the extra-embryonic parts; they are not directly involved in supplying cells for the embryo but break down yolk and transport it to the developing embryo.

Higher mammals—apart from the egg-laying mammals—do not have yolk in their eggs but, having passed through an evolutionary stage of animals with yolky eggs, retain, particularly in gastrulation, features common to reptiles (and birds, which also had reptilian ancestors). As a result, at the end of cleavage the formative cells of the embryo—the cells that will actually build the body of the animal—are arranged in the form of a disk over a cavity that takes the place of the yolk of the reptilian ancestors of mammals. Within the disk of cells a primitive streak develops, and the three germinal layers are formed much as in many reptiles and birds.

Gastrulation and the formation of the three germinal layers is the beginning of the subdivision of the mass of embryonic cells produced by cleavage. The cells then begin to change and diversify under the direction of the genes. The genes brought in by the sperm exert control for the first time; during cleavage all processes seem to be under control of the maternal genes. In cases of hybridization, in which individuals from different species produce offspring, the influence of the sperm is first apparent at gastrulation: paternal characteristics may appear at this stage; or the embryo may stop developing and die if the paternal genes are incompatible with the egg (as is the case in hybridization between species distantly related).

The diversification of cells in the embryo progresses rapidly during and after gastrulation. The visible effect is that the germinal layers become further subdivided into aggregations of cells that assume the rudimentary form of various organs and organ systems of the embryo. Thus the period of gastrulation is followed by the period of organ formation, or organogenesis.

Embryonic Adaptations

Throughout its development the embryo requires a steady supply of nourishment and oxygen and a means for disposal of wastes. These needs are met in various ways, depending in particular on (1) whether the eggs develop externally (oviparity), are retained in the maternal body until ready to hatch (ovoviviparity), or are carried in the maternal body to a later stage (viviparity); and (2) the length of embryonic development.

Adaptations in Animals other than Mammals

Eggs of many marine invertebrates are discharged directly into water, and the period of development before the larva emerges is relatively brief. Oxygen diffuses easily into the small eggs, and nourishment is provided by a moderate amount of yolk. During cleavage the yolk is distributed to all the blastomeres. Much of the nourishment in the egg is stored as animal starch, or glycogen, which is almost completely used by the time the larva emerges from the egg. A small amount of water and inorganic salts are taken in by the embryo from surrounding seawater. Eggs developing in freshwater carry their own supply of necessary amounts of certain salts that are not present in sufficient quantities in the environment. Products of metabolism—especially carbon dioxide and nitrogenous wastes in the form of ammonia—diffuse out from small embryos developing in water.

The eggs of terrestrial animals must overcome the hazard of drying. In certain species this danger is avoided because the animal returns to water to breed, such as frogs and salamanders. Some groups of insects (*e.g.*, dragonflies, mayflies, and mosquitoes) also lay eggs in water, and the larvae are aquatic. Eggs of other animals (*e.g.*, snails, earthworms) are laid in moist earth and thus are protected from drying up. In terms of evolution, however, a decisive solution to the problem of development on land was arrived at by most insects and by reptiles and birds, which developed eggs with a shell impermeable to water or, at least, resistant to rapid evaporation. The shells of bird and insect eggs, while restricting evaporation of water, allow oxygen to diffuse into the egg and carbon dioxide to diffuse out. Apart from gas exchange, the eggs constitute closed systems, which give nothing to the outside and require nothing from it. Such eggs are called cleidoic. Because the products of nitrogen metabolism in cleidoic eggs cannot pass through the eggshell, animals (birds and insects) have had to evolve a method of storing wastes in the form of uric acid, which, since it is insoluble, is nontoxic to the embryo.

After a short period of development in the egg, the emerging young animal has to fend for itself, unless there is some form of parental care. Exposure to the external environment at a tender age results frequently in loss of life, a hazard met by many animals through an increase in the supply of nourishment within the egg, thus allowing the young to attain a greater size and development. This tendency to produce large yolky eggs has been achieved independently in different evolutionary lines: in octopuses and squids among the mollusks, in sharks among the fishes, and in reptiles and birds among the terrestrial vertebrates.

As has been indicated, cleavage is incomplete in eggs with large amounts of yolk. Although some yolk platelets may be enclosed in the formative cells of the embryo, the bulk of the yolk remains an uncleaved mass, overgrown and surrounded by the cellular part of the embryo. In such cases a membranous bag, or yolk sac, is formed and remains connected to the embryo by a narrow stalk (the evolutionary precursor of the umbilical cord of mammals). The cellular layers surrounding the yolk sac and forming its walls may consist of all three germinal layers (in reptiles and birds), so that the yolk virtually comes to lie inside an extension of the gut of the embryo; or (in bony fishes) the yolk sac may be enclosed in layers of ectoderm and mesoderm. In either case a network of blood vessels develops in the walls of the yolk sac and transports the yolk products to the embryo. As the yolk is broken down and utilized, the yolk sac shrinks and is eventually drawn into the body of the embryo. In addition to the yolk sac, extra-embryonic parts are also encountered in the form of embryonic membranes, which are found in higher vertebrates and in insects. Vertebrates have three embryonic membranes: the amnion, the chorion, and the allantois.

In reptiles, birds, and mammals, folds develop on the surface of the yolk sac just outside and around the body of the embryo proper. These folds, consisting of extra-embryonic ectoderm and extra-embryonic mesoderm, rise up and fuse dorsally, enclosing the embryo in a double-lined, fluid-filled chamber known as the amniotic cavity. The inner lining of the fold becomes the amnion, and the outer becomes the chorion, which ultimately surrounds the entire embryo. The amniotic fluid protects the embryo from drying, prevents the adhesion of the embryo to the inner surface of the shell, and provides the embryo with efficient shock absorption against possible damaging jolts. (The aminion and chorion develop in the same way in insect embryos.) The third membrane, or allantois, is originally nothing more than the urinary bladder of the embryo. It is a saclike growth of the floor of the gut, into which nitrogenous wastes of the embryo are voided. It enlarges greatly during the course of development, eventually expanding between the amnion and chorion and also between the chorion and the yolk sac, to become the third embryonic membrane. In addition to providing storage space for the nitrogenous wastes of the embryo, the allantois takes up oxygen, which penetrates into the egg from the exterior, and delivers it, by way of a network of blood vessels, to the embryo.

Adaptations in Mammals

At some early stage during the evolution of viviparous mammals, eggs came to be retained in the oviducts of the mother. The embryo then was provided with nourishment from fluids in the oviduct; the yolk, which became redundant, gradually ceased to be provided, and the eggs became oligolecithal. The eggshell, present in reptiles, was no longer needed and eventually disappeared, as did the white of the egg. The chorion, however, remained as the most external coat of the developing embryo through which nourishment reaches the embryo. It acquired the ability to adhere closely to the walls of the uterus (which was what that part of the oviduct holding the embryo had become) and became the so-called trophoblast. The blood-vessel network of the underlying allantois conveys nutrients that diffuse through the trophoblast to the body of the embryo proper. These modifications gave rise to a new organ, the placenta, formed from tissues of both the mother and the embryo: the uterine wall with its blood vessels provided by the mother; the trophoblast and allantois—and in some mammals also the yolk sac—with their blood vessels provided by the embryo.

The overall development of placental mammals as a result of these changes is profoundly different from that of their ancestors, the reptiles, and proceeds in the following way: the tiny yolkless egg is fertilized in the upper portion of the oviduct by sperm received from the male in the process of coupling (coitus); cleavage starts as the egg is propelled slowly down the oviduct by action of cilia in the oviduct lining. At the end of cleavage a solid ball of cells called a morula is produced. The surface cells of the morula become the trophoblast and the inner cell mass gives rise to the embryo (the formative cells) and also its yolk sac, amnion, and allantois. A cavity appears within the morula, converting it into a hollow embryo, called the blastocyst. This cavity resembles the blastocoel but, in fact, is analogous to the yolk sac of meroblastic eggs, except that there is no yolk and the cavity is filled with fluid. At the blastocyst stage, the embryo enters the uterus and attaches itself to the uterine wall. This attachment, or implantation, a crucial step in the development of a mammal, is attained through the action of the trophoblast, which forms extensions, known as villi, that penetrate the uterine wall. In higher placental mammals, the lining of the uterine wall and, in varying degrees, the underlying tissues as well are partially destroyed, resulting in a closer relationship between the blood supplies of the mother and the embryo. Indeed, in man and in some rodents, the blastocyst sinks completely into the uterine wall and becomes surrounded by uterine tissue.

While implantation takes place, the formative cells arrange themselves in the form of a disk under the trophoblast. In the disk, the germinal layers develop much as in birds, with the formation of a primitive streak and migration of the chordamesoderm into a deeper layer. A layer of endoderm is formed adjoining the cavity of the blastocyst, and an amniotic cavity develops, enclosing the embryo; in lower placental mammals, the allantois also develops. The embryo proper, lying in the amniotic cavity, is connected to the extra-embryonic parts by the umbilical cord. The umbilical cord lengthens greatly during later development. In higher mammals, the cavity of the allantois is reduced, but the allantoic blood vessels become well developed and extend through the umbilical cord, connecting the embryo to the placenta. The blood that circulates in the placenta brings oxygen and nutrients from the maternal blood to the embryo and carries away carbon dioxide and other waste products from the embryo to the maternal blood for disposal by the maternal body.

Although tissues of maternal and embryonic origin are closely apposed in the placenta, there is little actual mingling of the tissues. Despite an occasional penetration of an embryo cell into the mother and vice versa, there is a placental barrier between the two tissues. The blood circulation of the mother is at all times completely separated from that of the embryo and its extra-embryonic parts. The placental barrier, however, does allow molecules of various substances to pass through; such differential permeability is indeed necessary if the embryo is to obtain nourishment. The permeability of the placental barrier differs in different animals; thus antibodies, which are protein molecules, may penetrate the placental barrier in man but not in cattle.

The maintenance of the fetus—as the more advanced embryo of a mammal is called—in the uterus is under hormonal control. In the initial stages of pregnancy, the continued existence of the embryo in the uterus depends on the hormone progesterone, which is secreted by the corpora lutea, "yellow bodies," that develop in the ovary after an egg has been released.

At birth the fetal parts of the placenta separate from the maternal parts. Contraction of the uterine wall first releases the fetus from the uterus; the fetal parts of the placenta (the afterbirth) follow. In certain cases of intimate connection between fetal and maternal tissues, the maternal tissues are torn, and birth is accompanied by profuse bleeding.

Organ Formation

Primary Organ Rudiments

Immediately after gastrulation—and sometimes even while gastrulation is underway—the germinal layers begin subdividing into regions that will give rise to various parts of the body. Subdivision proceeds in stages: initially a mass of cells is set aside for an organ system (for the alimentary canal, for instance) and subsequently further subdivided into the rudiments of various parts of the organ system, such as the liver, stomach, and intestines. The initially formed larger units are referred to as primary organ rudiments; those they later give rise to, as secondary organ rudiments.

Differentiation of the Germinal Layers

The type of organ rudiment produced depends on the organization of the body in any particular group in the animal kingdom. In the vertebrates the earliest subdivision within a germinal layer is the segregation within the chordamesodermal mantle of the rudiment of the notochord from the rest of the mesoderm. During gastrulation the material of the notochord comes to lie middorsally in the roof of the archenteron. It separates by longitudinal crevices from the chordamesodermal mantle lying to the left and right. The material of the notochord then rounds off and becomes a rod-shaped strand of cells immediately under the dorsal ectoderm, stretching from the blastopore toward the anterior end of the embryo, to the midbrain level. In front of the tip of the notochord, there remains a thin sheet of prechordal mesoderm.

The mesodermal layer adjoining the notochord becomes thickened and, by transverse crevices, subdivided into sections called somites. The somites, which later give rise to the segmented body muscles and the vertebral column, are the basis of the segmented organization typical of vertebrates (seen especially in the lower fishlike forms but also in the embryos of higher vertebrates). The lateral and ventral mesoderm, which remains unsegmented, is called the lateral plate. The somites remain connected to the lateral plate by stalks of somites that play a particular role in the development of the excretory (nephric) system in vertebrates; for this reason they are called nephrotomes. Rather early the mesodermal mantle splits into two layers, the outer parietal (somatic) layer and the inner visceral (splanchnic) layer, separated by a narrow cavity that will expand later to form the coelomic, or secondary, body cavity. The coelomic cavity extends initially through the nephrotomes into the somites; in the somites it is eventually obliterated. Endoderm completely surrounds the lumen of the archenteron (when present) and produces the cavity of the alimentary canal. If no archenteric cavity is formed during gastrulation, the cavity of the alimentary canal is formed by the separation of cells in the middle of the mass of endoderm (as in bony fishes) or by folding of the sheet of endoderm. The endodermal gut sooner or later acquires an extended anterior part called the foregut and a narrower and more elongated posterior part, the hindgut. Characteristic of chordates is the development of the nervous system from a part of ectoderm lying originally on the dorsal side of the embryo, above the notochord and the somites. This part of the ectodermal layer thickens and becomes the neural plate, whose edges rise as neural folds that converge toward the midline, fuse together, and form the neural tube. In vertebrates the neural tube lies immediately above the notochord and extends beyond its anterior tip. The neural tube is the rudiment of the brain and spinal cord; its lumen gives rise to the cavities, or ventricles, of the brain and to the central canal of the spinal cord. The remainder of the ectoderm closes over the neural tube and becomes, in the main, the covering layer (epithelium) of the animal's skin (epidermis). As the neural tube detaches itself from the overlying ectoderm, groups of cells pinch off and form the neural crest, which plays an important role in the development of, among other things, the segmental nerves of the brain and spinal cord.

In developing the primary organ rudiments, the embryo acquires a definite organization clearly recognizable as that of a chordate animal. Similar processes, which occur in the development of other animals, establish the basic organization of an annelid, a mollusk, or an arthropod.

Embryonic Induction

The organization of the embryo as a whole appears to be determined to a large extent during gastrulation, by which process different regions of the blastoderm are displaced and brought into new spatial relationships to each other. Groups of cells that were distant from each other in the blastula come into close contact, which increases possibilities for interaction between materials of different origin. In the development of vertebrates in particular, the sliding of cells (presumptive mesoderm) into the interior and their placement on the dorsal side of the archenteron (in the archenteric "roof"), in immediate contact with the overlying ectoderm, is of major importance in development and subsequent differentiation. Experiments have shown that, at the start of gastrulation, ectoderm is incapable of progressive development of any kind; that only after invagination, with chordamesoderm lying directly underneath it, does ectoderm acquire the ability for progressive development. The dorsal mesoderm, which later differentiates into notochord, prechordal mesoderm, and somites, causes the overlying ectoderm to differentiate as neural plate. Lateral mesoderm causes overlying ectoderm to differentiate as skin. The influence exercised by parts of the embryo, which causes groups of cells to proceed along a particular path of development, is called embryonic induction. Though induction requires that the interacting parts come into close proximity, actual contact is not necessary. The inducing influence—whatever it might be—is a diffusible substance emitted by the activating cells (the inductor). The inducing substance of the mesoderm is a large molecule, probably a protein or a nucleoprotein, which presumably penetrates reacting cells, though direct and unequivocal proof of such penetration is still unavailable. Inducing substances are active on vertebrates belonging to many different classes; *e.g.,* inductions of primary organs have been obtained by transplanting mammalian tissues into frog embryos or by transplanting tissues of a chick embryo into the embryo of a rabbit.

Induction is responsible not only for the subdivision of ectoderm into neural plate and epidermis but also for the development of a large number of organ rudiments in vertebrates. The notochord is a source of induction for the development of the adjoining somites and nephrotomes; the latter appear jointly to induce development of limb rudiments from the lateral plate mesoderm.

Since the results of induction are different for different organ rudiments, it must be presumed that there exist inducing substances with specific action, at least to a certain extent; thus, the lateral mesoderm induces differentiation of the skin but not neural plate from the very same kind of ectoderm. The number of inducing substances need not, however, be the same as the number of different kinds of tissues and organs, since certain differentiations could possibly be induced by a combination of two or more inducing substances, or the same inducing substance might have different effects on different tissues. It has been suggested that the regional organization of the entire vertebrate body could be controlled by the graded distribution of only two inducing substances—provisionally named the neuralizing substance and the mesodermalizing substance—along the length of the embryo. The neuralizing substance, concentrated at the anterior end, gradually decreases toward the posterior end; the mesodermalizing substance, on the other hand, is concentrated at the posterior end and decreases toward the anterior end. The differentiation of induced structures depends on the relative amounts of the two inducing substances at any given point in the embryo. Acting alone, the neuralizing substance induces only nervous tissue, which takes the form of the forebrain, and the mesodermalizing substance induces only mesodermal structures (*e.g.,* somites, notochord).

In the amphibian embryo, induction appears to have its primary source in the dorsal lip of the blastopore, which eventually gives rise to the notochord and adjoining somites. Induction by the notochord and somites is responsible for the development of the neural plate in the ectoderm, of lateral and ventral parts of the mesodermal mantle, and of the lumen of the alimentary canal in the endoderm. The dorsal lip of the blastopore for this reason has been called the primary organizer. In higher vertebrates, in which gastrulation occurs through the medium of a primitive streak, the anterior end of the streak and the Hensen's node have properties similar to those of a primary organizer. Organization centres have been found, or suspected, in embryos of animals belonging to a few other groups, in particular the insects and sea urchins, but the interpretation of the experimental results in these animals is less satisfactory than in the case of vertebrates.

The concept of an organization centre suggests that a part of the embryo differs from the rest of the embryonic tissues in being more active. The more active parts of the embryo (and also of animals in later stages of development) are particularly sensitive to certain noxious influences in their environment. If an embryo is deprived of oxygen or subjected to weak concentrations of poisons, the first parts to suffer are the most morphogenetically active ones. In vertebrate embryos the anterior end of the head is most sensitive. Early sea-urchin embryos have two centres of maximal sensitivity: one at the animal pole and the other at the vegetal pole. The damage done by noxious influences may result in actual breakdown of cells in a region of maximal sensitivity and may also lead to a depression of the developmental potential of the cells. Thus, the graded distribution of certain physiological properties appears to play a part in morphogenetic processes: physiological gradients are in fact also morphogenetic gradients.

Gradients in the embryo can be used to control development to a certain extent, by exposing the embryo to influences that, while reaching all parts, have a local effect as the result of differences in sensitivity. Disturbances of normal development often are the result of disruptions of gradients.

Organogenesis and Histogenesis

The primary organ rudiments continue to give rise to the rudiments of the various organs of the fully developed animal in a process called organogenesis. The formation of organs, even those of diverse function, shares some common features. As the organs form, so do their component tissues, in a process termed histogenesis.

A germinal layer, as the name implies, is a sheet of cells. An organ rudiment may be formed and separated from such a sheet in several ways. A groove, or fold, may appear within the layer, become closed into a tube, and then separated from the original layer. A tube once formed may be subdivided into sections by constrictions and dilations of the tube at certain points. This is the way the nervous system rudiment is formed in vertebrates.

Alternatively, the germinal layer may produce a round depression, or pocket. The pocket may then separate from the layer as a vesicle, or it may elongate and branch at the tip while still connected with the layer. The latter method is common in the development of various glands and also the lungs in vertebrates.

Still another method of rudiment formation in a germinal layer is by the development of local thickenings, elongated or round, and detachment from the epithelial sheet. If a lumen appears later within such a body, the result may be the same as that achieved by folding—that is, a tube or vesicle may be formed. Indeed, the same sort of organ may develop even in related animals in either of these ways. The epithelial layer may further be cut up into segments, with the layer losing continuity, as in the formation of somites in vertebrates or similar mesodermal blocks in segmented invertebrates (*e.g.*, annelids and arthropods).

Lastly, the cells of a germinal layer may give up their connection to each other and become a mass of loose, freely moving cells called embryonic mesenchyme. This mass gives rise to various forms of connective tissue but may also condense into more solid structures, including parts of the skeleton and the muscles.

Many organs are comprised of all three germinal layers. It is very common for glands, for instance, to derive their lining from an ectodermal or endodermal epithelium and their connective tissue (sometimes in the form of a capsule) from mesenchyme of mesodermal origin. Parts of ectoderm and endoderm cooperate also in the development of the lining of the alimentary canal, and mesoderm provides the connective tissue and muscular sheath of the canal.

The development of organs of the body are dealt with according to the germinal layer that contributes the most important part, and only the development of vertebrate organs is considered.

Ectodermal Derivatives

The Nervous System

The vertebrate nervous system develops from the neural plate—a thickened dorsal portion of the ectoderm—which forms a tube. From the very start the tube is wider anteriorly, the end that gives rise to the brain. The posterior part of the neural tube, which gives rise to the spinal cord, is narrower and stretches as the embryo lengthens. Stretching involves the head to only a very minor degree.

The Brain and Spinal Cord

Constrictions soon appear in the brain region of the neural tube, subdividing it into three parts, or brain vesicles, which undergo further transformations in the course of development. The most anterior of the primary brain vesicles, called the prosencephalon, gives rise to parts of the brain and the eye rudiments. The latter appear in a very early stage of development as lateral protrusions from the wall of the neural tube, which are constricted off from the remainder of the brain rudiment as the optic vesicles. The rest of the prosencephalon constricts further into two portions, an anterior one, or telencephalon, and a posterior one, or diencephalon. The telencephalon gives rise, in lower vertebrates, to the smell, or olfactory, centre; in higher vertebrates and man, it becomes the centre of mental activities. The diencephalon, with which the eye vesicles are connected, was presumably originally an optic centre, but it has acquired, in the course of evolution, a function of hormonal regulation. The floor of the diencephalon forms a funnel-shaped depression, the infundibulum, which becomes connected with the pituitary, or hypophysis, the most important gland of internal secretion (*i.e.*, endocrine gland) in vertebrates. Indeed, the posterior lobe of the hypophysis is actually derived from the floor of the diencephalon. Tissues of the infundibulum and the posterior lobe of the hypophysis produce certain hormones (oxytocin and vasopressin) and stimulate the production and release of other hormones from the anterior lobe of the hypophysis.

The second primary brain vesicle, the mesencephalon, gives rise to the midbrain, which, in higher vertebrates, takes part in coordinating visual and auditory stimuli.

The third primary brain vesicle, the rhombencephalon, is more elongated than the first two; it produces the metencephalon, which gives rise to the cerebellum with its hemispheres, and the myelencephalon, which becomes the medulla oblongata. The cerebellum acts as a balance and coordinating centre, and the medulla controls functions such as respiratory movements.

The cells constituting the wall of the neural tube and, later, of the brain and spinal cord become arranged in such a way that they point into the central cavity of the tube. The differentiation of nervous tissue involves many cells abandoning their connection to the inner surface of the neural tube and migrating outward, where they accumulate as a mantle. The first cells to migrate become the neurons, or nerve cells. They produce outgrowths called axons and dendrites, by which the cells of the nervous system establish communication with one another to form a functional network. Some of the outgrowths extend beyond the confines of the brain and spinal cord as components of nerves; they establish contact with peripheral organs, which thus fall under the control of the nervous system. Cells migrating from the inner surface of the neural tube later in development become astrocytes, which are the supporting elements of nerve tissue.

The fate of nerve cells is dependent largely on whether they succeed, directly or indirectly (through other neurons), in connecting with peripheral organs. Nerve cells that fail to establish connections die. Thus, if in early stages of embryonic development, some organ, a limb rudiment for instance, is surgically removed, the nerve cells in the centres supplying nerves to such an organ are reduced in number, and the corresponding nerves also diminish or disappear. On the other hand, if an organ is introduced by transplantation into a developing embryo, the organ will be supplied by nerves from a nerve centre in which the number of cells apparently increases; no additional cells are provided, but cells that would otherwise have degenerated remain active and differentiate into functional neurons, thus satisfying the demand created by the additional organ.

Nerves do not consist entirely of outgrowths of neurons located in the brain and spinal cord. Many components of nerves are outgrowths of neurons, the cell bodies of which are located in masses called ganglia; there are three main types of ganglia: spinal ganglia, cranial ganglia, and ganglia of the autonomous nervous system. The spinal ganglia are derived from cells of the neural crest—the loose mesenchyme-like tissue that remains between the neural tube and skin after separation of the two. Part of the cells of the neural crest in the region of the trunk and tail accumulate in seg-mental groups (corresponding to the mesodermal somites) and provide fibres to peripheral organs and to the spinal cord. These fibres constitute the sensory pathways in the spinal nerves. The motor components of the spinal nerves—fibres that activate muscles—are outgrowths of neurons lying in the spinal cord. The ganglia of the cranial nerves are produced only in part from cells of the neural crest; an additional component comes from the epidermis on the side of the head. Cells of the epidermal thickenings called placodes detach themselves and contribute to the formation of the cranial ganglia and thus of the cranial nerves.

The ganglia of the autonomous (sympathetic) nervous system are derived, as are the spinal gan-glia, from neural-crest cells, but, in this case, the cells migrate downward to form groups near the dorsal aorta, near the intestine, and even in the intestinal wall itself. The outgrowths of cells in these ganglia are the nerve fibres of the sympathetic nerves.

Major Sense Organs

The Eye

As has been pointed out, the rudiments of the eyes develop from optic vesicles, each of which re-mains connected to the brain by an eye stalk, which later serves as the pathway for the optic nerve. The optic vesicles extend laterally until they reach the skin, whereupon the outer surface caves in so that the vesicle becomes a double-walled optic cup. The thick inner layer of the optic cup gives rise to the sensory retina of the eye; the thinner outer layer becomes the pigment coat of the retina. The opening of the optic cup, wide at first, gradually becomes constricted to form the pupil, and the edges of the cup surrounding the pupil differentiate as the iris. The refractive system of the eye and, in particular, the lens of the eye are derived not from the cup but from the epidermis overlying the eye rudiment. When the optic vesicle touches the epidermis and caves in to produce the optic cup, the epidermis opposite the opening thickens and produces a spherical lens rudiment. The lens develops by an induction by the optic vesicle on the epidermis with which it comes in contact. A further influence emanating from the eye changes the epidermis remaining in place over the lens into a transparent area, the cornea. Influence of the optic cup on the surrounding mesenchyme causes the latter to produce a vascular layer around the retina and, outside of that, a tough fibrous or (in some animals) even a partly bony capsule called the sclera. Thus a complex interdependence of different materials produces the fully developed and functional vertebrate eye.

The Ear

The main part of the ear rudiment is derived from thickened epidermis adjoining the medulla. This area of the epidermis invaginates to produce the ear vesicle, which separates from the epidermis but remains closely apposed to the medulla. The ear vesicle becomes complexly folded to produce the labyrinth of the ear. Subsequently, a group of cells of the ear vesicle becomes detached and gives rise to the acoustic ganglion. Neurons of this ganglion become connected by their nerve fibres to the sensory cells in the labyrinth, on the one hand, and with the brain (the medulla), on the other. The ear vesicle, acting on the surrounding mesenchyme, induces the latter to aggregate around the labyrinth and form the ear capsule. Further parts with various origins are added to the ear: the middle ear, from a pharyngeal pouch and the associated skeleton, and the external ear (where present), from epidermis and dermis.

The Olfactory Organ

The olfactory organ develops from a thickening of the epidermis adjacent to the neural fold at the anterior end of the neural plate. This thickening is converted into a pocket or sac but does not lose connection with the exterior. The openings of the sac become the external nares, and the cavity of the sac becomes the nasal cavity. Some cells of the olfactory sac differentiate as sensory epithelium and produce nerve fibres entering the forebrain. In most fishes the olfactory sac does not communicate with the oral cavity; in lungfishes and in terrestrial vertebrates, however, canals develop from the olfactory sacs to the oral cavity, where they open by internal nares. A cartilaginous capsule forms around the olfactory organ from cells believed to have been derived from the walls of the sac itself, and thus it is ectodermal in origin.

Gustatory and other Organs

Gustatory organs in the form of taste buds develop as local differentiations of the lining of the oral cavity but also, in fishes, in the skin epidermis. They are supplied with nerve endings, as are several other sensory bodies scattered among the tissues and organs of the developing body.

The Epidermis and its Outgrowths

The major part of the ectodermal epithelium covering the body gives rise to the epidermis of the skin. In fishes and aquatic larvae of amphibians, the many-layered epidermis is provided with unicellular mucous glands. In terrestrial vertebrates, however, the epidermis becomes keratinized; *i.e.,* the outer layers of cells produce keratin, a protein that is hardened and is impermeable to water. During the process of keratinization, many cell components degenerate and the cells die; the layer of keratinized cells is therefore shed from time to time. In reptiles the shedding may take the form of a molt in which the animal literally crawls out of its own skin. It is less well known that frogs and toads also molt, shedding the surface keratinized layer of their skin (which is usually eaten by the animal). In birds and mammals, keratinized cells are shed in pieces that are sloughed off, rather than in extensive layers. In many vertebrates local thickenings of the keratinized layer appear in the form of claws, hooves, nails, and horns.

The epidermis is only the superficial layer of the skin, which is reinforced by the dermis, a connective tissue layer of a much greater thickness. The cells of the dermis are derived from mesoderm and neural-crest cells. In particular the pigment cells found in the dermis of fishes, amphibians, and reptiles are of neural-crest origin. The pigment in the skin of birds and mammals (and also in hairs and feathers) is also produced by neural-crest cells, but in these animals the pigment cells penetrate into the epidermis or deposit their pigment granules there.

The structure of the skin is further complicated by the development of hairs and feathers, on the one hand, and of skin glands, on the other. Hairs and feathers develop from a somewhat similar kind of rudiment. The development starts with a local thickening of the epidermal layer, beneath which a group of mesenchyme cells accumulate. In the case of hairs, the epidermal thickening proliferates downward and forms the root of the hair, from which the shaft then grows outward, emerging on the surface of the skin. In the case of feathers, the epidermal thickening bulges outward to form a hollow fingerlike protrusion with a connective tissue core. Secondarily, the shaft of the feather branches characteristically to produce barbs and barbules. In both cases, however, the final structure—shaft of the hair and shaft barbs and barbules of the feather—consists of keratinized and, thus, dead cells.

The skin of amphibians and mammals (but not of birds and reptiles) is provided with numerous skin glands, which develop as ingrowths from the epidermis. A peculiar type of skin gland is the mammary gland of placental mammals. In the first stage of development, mammary-gland rudiments resemble hair rudiments; they are thickenings of the epidermis, with condensed mesenchyme on their inner surfaces. In some mammals (rabbit, man) two continuous epidermal thickenings called mammary lines stretch along either side of the belly of the embryo. Parts of the line corresponding in number and position to the future glands enlarge while the rest of the thickening disappears. The initial thickenings proliferate inward and produce a system of ramified cords, solid at first but hollowed out later, which become the lactiferous, or milk-bearing, ducts of the gland. Further branching at the tips of the ducts gives rise to smaller ducts and to the secretory end sacs, or alveoli, of the gland.

Mesodermal Derivatives

The Body Muscles and Axial Skeleton

The somites, formed in the early stages of development from the upper edges of the mesodermal mantle adjoining the notochord, are complex rudiments that subdivide and give rise to very diverse body structures. The coelomic cavity, present initially, becomes obliterated by the side-to-side flattening of the somites, so that the thinner, outer parietal layer of the somite comes in close contact with its thicker visceral layer. The visceral layer of the somite very early subdivides into two parts. The upper, dorsolateral part called the myotome remains compact, giving rise to the body muscles. The lower, medioventral part of the somite, called the sclerotome, breaks up into mesenchyme, which contributes to the axial skeleton of the embryo—that is, the vertebral column, ribs, and much of the skull. The parietal layer of the somite, at a later stage, is converted into mesenchyme that, together with components of the neural crest, gives rise to the dermis of the skin and, for this reason, is called the dermatome.

The cells of the myotome are elongated in a longitudinal direction and become differentiated as muscle fibres. The myotomes, originally situated dorsally, expand on either side, penetrating between the skin on the outside and the lateral plates of the mesoderm on the inside, until they meet midventrally; the whole body is thus enclosed in a layer of developing muscle. As the somites and myotomes are segmented, so are the muscles derived from them. Metamerism, or segmentation, a feature in the embryos of all vertebrates, remains preserved only in the adults of fishes and of terrestrial vertebrates that have elongated bodies (salamanders, snakes); it becomes largely erased in four-footed animals that depend on their limbs for locomotion.

The mesenchyme derived from the sclerotomes condenses as cartilage around the notochord and the spinal cord. It forms the cartilaginous vertebral column and ribs. In the head region it produces a part of the cartilaginous skull, mainly its posterior and ventral parts; anteriorly the somitic mesenchyme is supplemented by mesenchyme from the neural crest. Cartilaginous capsules of the olfactory organ and the ear fuse with the cartilaginous capsule surrounding the brain; to this complex are also added cartilages associated with the jaws and gill skeleton. Cartilage in the vertebral column and in the skull is replaced later in the bony fishes and in the terrestrial vertebrates by bone. At a still later stage, dermal bones are added, which, while they have no precursors in the cartilaginous skeleton, develop in the adjoining mesenchyme.

The Appendages: Tail and Limbs

The tail in vertebrates is a prolongation of the body beyond the anus. It develops in early stages from the tail bud, immediately dorsal to the blastopore. Material for the tip of the tail is situated slightly forward from the edge of the blastopore. The elongation of the back of the body is greater than that of the belly; as a result the tip of the tail bud is carried beyond the blastopore and thus beyond the anus, which, in the developed embryo, marks the position of the blastopore. The consequence is that a section of the dorsal surface of the embryo comes to lie on the ventral surface of the tail; *i.e.*, becomes inflected. The tail bud is formed from parts that have already been differentiated to a certain extent; prolongations of the neural tube and of the notochord are involved, and endoderm extends into the tail rudiment as the postanal gut, which, however, soon degenerates. The bud is also encased in ectodermal epidermis. In amphibians the somites of the tail are not derived from the chordamesodermal mantle but from the inflected posterior portion of the neural plate, which loses its nervous nature and becomes subdivided into segments corresponding to the somites of the trunk. In higher vertebrates the cells in the interior of the tail bud have an undifferentiated appearance and form a growth zone, at the expense of which parts of the tail (neural tube, notochord, somites) are extended backward as the tail elongates.

The paired limbs of vertebrates derive their first rudiments from the upper edge of the lateral plate mesoderm. The parietal layer becomes thickened, and cells escape from the epithelial arrangement and form a mesenchymal mass adjoining the ectodermal epithelium at the surface of the body. The ectodermal epithelium over the mass of mesenchyme likewise becomes thickened. In higher vertebrates, the accumulation of mesodermal cells and the thickening of the epidermis occur along the entire length of the trunk, from neck to anus, but in the middle of the trunk they soon disappear, and only the most anterior and the most posterior sections develop further into the rudiments of the forelimbs and hindlimbs, respectively. In fishes, the rudiments of the pectoral and pelvic fins are more extended anteroposteriorly in earlier than in final stages.

The mesodermal masses of the limb rudiments proliferate, and, covered with thickened epidermis, form on the surface of the body conical protrusions called the limb buds, which, once formed, possess all the materials necessary for limb development. Limb buds may be transplanted into various positions on the body or on the head and there develop into clearly recognizable limbs, conforming to their origin, whether a forelimb or hindlimb, a wing or a leg in birds. This specificity of the limb is carried by the mesodermal part of the rudiment, but a complex interaction between the mesodermal mesenchyme and the ectodermal epidermis is necessary for the normal development of the limb. In four-limbed vertebrates (tetrapods), the tips of the limb buds become flattened and broadened into hand or foot plates. The edge of the plate is indented, forming the rudiments of the digits. Meanwhile, local areas of the mesodermal mesenchyme in the interior of the limb rudiment condense; these are the rudiments of the various components of the limb skeleton. In fishes, small outgrowths from the myotomes enter the limb rudiment to form the muscles of the fins. In tetrapods, however, the limb muscles develop from the same mass of mesenchyme that gives rise to the skeleton. Thus the muscles of the body and the muscles of the limbs have different origins—the first develop from the myotomes (thus from the somites), and the second develop from the lateral plate mesoderm via the limb buds.

The nerves supplying the limbs grow into the limb rudiments from the spinal cord and the spinal ganglia. The nerves are guided in some way by the limb rudiments, for, if limb rudiments are displaced by transplantation to an abnormal position, the nerves still find their way and establish normal relationships to the limb muscles. Limb rudiments transplanted to sites very far from their normal positions induce local nerves to enter the limb, thereby making it motile.

Excretory Organs

The kidneys of vertebrates consist of a mass of tubules that develop from the stalks of somites called nephrotomes. In some primitive vertebrates such as cyclostomes, the nephrotome in each segment gives rise to only one tubule, but, in the great majority of vertebrates, mesenchyme from adjacent nephrotomes fuses into a common mass that differentiates into a number of nephric tubules irrespective of the original segmentation of the mesoderm. Under primitive conditions each tubule opens by a funnel (the nephrostome) into the coelomic cavity; the opposite ends of the tubules fuse to form the collecting ducts of the kidney. A collection of capillaries (the glomerulus) becomes associated with the nephric tubule, forming its filtration apparatus. The glomerulus may be situated in the coelomic cavity opposite the nephrostome or, in all the more advanced animals, intercalated into the nephric tubule, forming with the latter a renal corpuscle of the kidney. In adults of all vertebrates above the amphibians, the nephrostomes disappear (or are never formed), so that the tubule begins with the renal corpuscle. Parts of the kidney in vertebrates can be distinguished as the pronephros (most anteriorly, at the forelimb level), the mesonephros (in the midtrunk region), and the metanephros (in the pelvic region). The three sections of the kidney develop at different stages, starting with the pronephros and ending with the metanephros. In their morphology and mode of development, the anterior parts show more primitive conditions than the posterior ones. The pronephros, developing early in embryo formation, is the functional kidney of fish and amphibian larvae. Its collecting duct opens into the hindmost part of the intestine, called the cloaca, and later also serves as the collecting duct of the mesonephros. In reptiles, birds, and mammals, the pronephros is nonfunctional, although even in these animals its duct persists as the mesonephric duct. The mesonephros develops later and replaces the pronephros as the functional kidney of adult fishes and amphibians and of the embryos of reptiles, birds, and mammals. The tubules of the mesonephros link up with the duct derived from the pronephros. The pronephric duct in fact stimulates the development of mesonephric tubules, and, in its absence, the mesonephros does not develop at all.

The metanephros is found only in reptiles, birds, and mammals. It replaces the mesonephros of the early embryonic stages and continues as the functional kidney in the postembryonic and adult life of these animals. The metanephros develops from mesenchyme derived from the nephrotomes of the posterior part of the trunk and lying dorsal to the mesonephric duct. The actual differentiation is initiated by a dorsal outgrowth of the mesonephric duct, called the ureteric bud. The ureteric bud grows in the direction of the mesenchyme and becomes the ureter. Having penetrated the mass of mesenchyme, it starts to branch, producing the collecting tubules of the kidney; the mesenchyme, meanwhile, in response to the influence of the duct and its branches, aggregates to form the excretory tubules of the kidney. The influence of the ureter is indispensable for the development of the metanephric excretory tubules, for, if the ureter fails to develop or, in its outgrowth, stops short of reaching the kidney-producing mesenchyme, no kidney develops.

Circulatory Organs

The rudiment of the heart in vertebrates develops from the ventral edges of the mesodermal mantle in the anterior part of the body, immediately adjoining the pharyngeal region. A group of mesodermal cells breaks away from the ventral edge of the lateral plate, takes a position just underneath the pharyngeal endoderm, and becomes arranged in the form of a thin-walled tube, which will become the endocardium, or lining of the heart. In vertebrates with complete cleavage, the endocardial tube is single and medial from its start. In higher vertebrates with meroblastic cleavage—reptiles, birds, and mammals—the embryo in early stages of development is flattened out on the surface of the yolk sac; therefore, what are morphologically the ventral edges of the mesodermal mantle lie far apart on the perimeter of the blastodisc. As a result of this arrangement, two endocardial tubes are formed, one on either side of the embryo. Subsequently, when the embryo becomes separated from the yolk sac, the two endocardial tubes meet in the midline ventral to the pharynx and fuse, producing a single heart rudiment. After the formation of the endocardium, or the lining of the heart, the coelomic cavity in the lateral plate mesoderm adjoining the heart rudiment expands slightly and envelops the endocardial tube or tubes. The heart muscle layer, or myocardium, develops from the visceral (splanchnic) layer of the lateral plate that is in contact with the endocardial tube; the parietal (somatic) layer of the lateral plate forms the pericardium, or covering of the heart. The portion of the coelom surrounding the heart becomes separated from the rest of the body cavity and develops into the pericardial cavity.

The endocardial tube branches anteriorly into two tubes, the ventral aortas; a similar branching of the endocardial tube posteriorly forms the two vitelline veins, which carry blood from the midgut endoderm or from the yolk sac (when present) to the heart.

In its earliest development, the heart rudiment shows a degree of dependence on the adjoining endoderm. The whole of the endoderm can be removed in newt embryos in the neural-tube stage. In such endodermless embryos, the heart fails to develop, even though the mesoderm destined to form the heart rudiment is left intact.

The heart is initially a straight tube stretching in an anteroposterior direction. Rather early in development, however, it becomes twisted in a characteristic way and subdivided into four main parts: the most posterior, the sinus venosus; the atrium, which comes to lie at the anteriorly directed bend of the tube; the ventricle, occupying the apex of the posteroventrally directed inflexion; and, most anteriorly, the conus arteriosus. In the course of development in the more advanced vertebrates, the atrium and ventricle become partially or completely subdivided into right and left halves. In amphibians, only the atrium is separated into two halves, by a partition starting from the posterior end. In reptiles, a partition separates the atria and part of the ventricle. In birds and mammals, the subdivision of the heart is complete, with two atria and two ventricles.

The complete subdivision of the heart is important for separating the pulmonary, or lung, blood supply from the general body circulation. But, if this separation developed early in the embryo, it would create difficulties, since the lungs of the embryo are not functional; the enrichment of the blood with oxygen occurs instead in the placenta. The partition between the atria in mammalian embryos remains incomplete, so that blood returning from the body and from the placenta enters into the right half of the heart but is shunted (through the interatrial foramen) into the left half of the heart and thence again into general circulation. At birth, however, the interatrial foramen is closed by a membraneous flap, and oxygen-depleted blood from the body enters the right atrium, is channelled into the right ventricle, and thence to the lungs for oxygenating.

In an adult vertebrate, blood vessels extend to all parts of the body. It would seem that channels for the supply of blood are provided in proportion to the local demand of the tissues; progressively developing organs or parts with particularly intensified function always receive an increased blood supply. The rudiments of blood vessels are always aggregations of mesenchyme cells. In any blood vessel the endothelial tube is formed first, and the muscular and elastic layers are added later.

The main blood channels in vertebrates develop in certain favoured situations; namely (1) between the endoderm and lateral plate mesoderm; (2) around the kidneys, especially the pronephros and mesonephros; and (3) in connection with the heart, which is a special case of the first category.

From the paired forward extensions from the heart, the ventral aortas, loops develop between the pharyngeal clefts. These are the aortic arches, which served originally to supply blood to the gills in aquatic vertebrates. The arches are laid down in all vertebrates, six or more being found in cyclostomes and fishes; six are present in the embryos of tetrapods, but the first two are degenerate. The arches of the third pair develop as the carotid arteries, supplying blood to the head. Those of the fourth pair (and, exceptionally, in urodeles also the fifth) join dorsally to form the dorsal aorta, providing blood to most of the body. These are the systemic arches. The arches of the sixth pair are the pulmonary arches; in embryos they carry blood to the dorsal aorta, as well as to the lungs, but in fully developed amniotes (reptiles, birds, and mammals), they carry blood only to the lungs.

The paired posterior extensions of the heart of the early embryo are the vitelline veins, whose branches spread out between the lateral plate mesoderm and the endoderm, especially the endoderm of the yolk sac, when present. On their way to the heart, the vitelline veins pass through the liver and break up into a system of small channels—the hepatic sinusoids. Parts of the vitelline veins lying posterior to the liver become the hepatic portal veins, which carry blood from the intestine to the liver; the parts of the vitelline veins anterior to the liver become the hepatic veins, which carry blood from the liver to the sinus venosus in lower vertebrates (anamniotes), but become the anterior section of the postcaval vein in amniotes.

Whereas the vitelline veins and, later, the hepatic portal vein carry blood from the endodermal parts of the embryo and from the yolk sac to the heart, the blood from the mesodermal and ectodermal parts is returned to the heart through a system of cardinal veins. These latter veins start their development in the form of an irregular sinus around the pronephros, connected by the common cardinal veins (ducts of Cuvier), on either side, to the sinus venosus. Extensions anteriorly and posteriorly give rise to the precardinal and postcardinal veins, respectively. The postcaval vein, present in terrestrial vertebrates, is a late acquisition, both in evolution and in embryogenesis; it is a result of the intercommunication of several venous channels, including the anterior portion of the vitelline veins.

The first blood cells in vertebrate embryos form in association with the intestinal endoderm on the yolk sac. Groups of mesoderm cells derived from the splanchnic layer of the lateral plate (extra-embryonically in cases in which a yolk sac is present) become so-called blood islands, which are particularly conspicuous on the yolk sac of bird embryos (in the area vasculosa). In bird's eggs, the internal cells of the blood island start producing hemoglobin (gas-carrying component of blood) and become the first red blood cells (erythrocytes) as early as the second day of incubation. The outer cells of the blood islands develop into an endothelial layer and form a network of blood vessels covering part of the surface of the yolk sac. The network acquires a connection to the vitelline veins and vitelline arteries (the latter being branches of the dorsal aorta); thus the blood corpuscles formed in the blood islands can enter the general blood circulation.

At later stages of embryogenesis, blood-cell formation shifts from the blood islands to the liver and, still later, to the bone marrow.

The lymphatic system, in a manner similar to the blood vessels, develops by the local aggregation of connective tissue to form lymphatic vessels.

Reproductive Organs

In considering the development of reproductive organs, distinctions must be made between: (1) the origin of sex cells (gametes), (2) the origin and differentiation of the sex glands, or gonads (ovaries and testes), and (3) the origin and development of the supporting parts of the reproductive system (*e.g.*, genital ducts, copulatory organs).

The germ (germinal) cells, which eventually give rise to the gametes, are often segregated from the somatic, or body, cells at a very early stage—during cleavage and before the subdivision of the embryo into ectoderm, mesoderm, and endoderm. In the invertebrate nematodes, the very first of these primordial germ cells is identifiable after as few as five divisions of the egg cell. The germ cell retains the large chromosomes present in the fertilized egg; in the somatic cells the chromosomes become fragmented. Subsequently, the single germ cell gives rise, by mitotic divisions, to all the gametes in the gonad.

In vertebrates, primordial germ cells arise outside the gonads, but they cannot be distinguished in early cleavage stages. In amphibians, cytoplasm at the vegetal pole, rich in ribonucleic acids, becomes incorporated into a number of cells, which, during cleavage and gastrulation, lie among the yolky endoderm cells. Later they migrate into the mesodermal layer and become incorporated into the rudiments of the gonads. In higher vertebrates, primordial germ cells can be recognized in the extra-embryonic endoderm of the yolk sac. In mammals, these cells subsequently migrate into the mesoderm and are located in the gonad rudiments. The mouse embryo, for example, originally has fewer than 100 primary germ cells; during their migration, however, their numbers increase as a result of repeated divisions, to 5,000 or more in the gonads.

Although the primordial germ cells either may appear before the separation of germinal layers or be found originally in the endoderm, the gonads are invariably of mesodermal origin. In vertebrates, the first trace of gonad development is a thickening of the coelomic lining on either side of the dorsal mesentery and medial to the kidney rudiments. The thickening, elongated anteroposteriorly, is known as the germinal ridge. The ridge protrudes into the coelomic cavity, and the fold of thickened epithelium becomes filled with mesenchyme. At this stage the primordial germ cells invade the rudiments of the gonads and become associated with the somatic cells of the germinal ridge. In the functionally differentiated gonads, only the actual gametes and their predecessors (spermatogonia and oogonia) are derived from the primary germ cells; the supporting cells are somatic cells of local mesodermal origin. In the ovaries, the follicle cells surrounding and nourishing the young egg cells (oocytes) are of somatic origin, as are also the connective tissue and blood vessels of the gonad. In the testes, supporting elements called Sertoli cells are somatic cells, as are the interstitial cells, which are scattered between the sperm-carrying tubules of the testes and believed to be the source of male hormones.

In the early stages of their development—even while the gonad rudiment is being invaded by primordial germ cells—the female and male gonads are in an indifferent stage. Only later does tissue differentiation of the gonads begin and male or female gonadal development proceed.

The genital ducts, by which the eggs and sperm are carried away from the gonads, are, in vertebrates, linked with the excretory system. In the male, the seminiferous tubules connect with the nephric tubules of the mesonephros, and the sperm are carried to the exterior by way of the mesonephric duct. In males of lower vertebrates, the mesonephric duct thus serves as a channel both for urine and for sex cells. In amniotes the development of the metanephros as the urine excreting organ has freed the mesonephric duct to carry products associated only with reproduction. In the female, a separate duct, the paramesonephric duct (Müllerian duct), develops beside the mesonephric duct. At its anterior end it utilizes the funnels of the pronephric tubules as its entrance (ostium). The paramesonephric duct develops initially in both female and male embryos. The ducts remain in an indifferent stage longer than the gonads. Eventually the sex hormones produced by the differentiating gonads cause a corresponding differentiation of the ducts. The mesonephric ducts, which become reduced in female embryos, remain in male embryos as ducts for conveying sperm (ductus deferens). The paramesonephric ducts, on the other hand, degenerate in male embryos but become the oviducts in female embryos. In mammals, the terminal portions of the paired oviducts differentiate as two uteri, which, in primates and man, fuse to form a single uterus.

In all terrestrial vertebrates except the placental mammals, the genital ducts, as well as the ducts of excretory organs, open into the cloaca. In mammals, however, the cloaca becomes subdivided into a dorsal part, which conveys the feces, and a ventral part, which receives excretory and genital products. In male mammals the excretory and genital ducts remain connected, having the urethra as their common outlet; in females the urethra serves only for the passage of urine and the uterus opens separately by means of the vagina. In nearly all vertebrates, the male nephric duct is utilized in some degree for the conduction of sperm.

Copulatory organs have developed independently in several groups of vertebrates having internal fertilization. The penis in mammals develops from an outgrowth called the genital tubercle, located at the anterior edge of the urinogenital orifice. The tubercle is laid down in a similar way in embryos of both sexes, and the region of the urinogenital orifice remains in an indifferent state even longer than do the genital ducts. In a comparatively late stage of embryonic life the genital tubercle of male embryos encloses the urethral canal and becomes the penis; in female embryos it remains small and becomes the clitoris.

Endodermal Derivatives

The Alimentary Canal

The alimentary canal is the chief organ developing from endoderm. The way it forms depends on the type of egg cleavage. In eggs with holoblastic (complete) cleavage, after gastrulation the invaginated mass of endoderm lines the archenteron, the cavity of which becomes the alimentary canal, or gut. In eggs with meroblastic (partial) cleavage—and also in mammals (despite their complete cleavage)—the endoderm is produced in the form of a sheet lying flat over the yolk-sac cavity. Subsequently, folds of endoderm and splanchnic mesoderm appear—first anteriorly, then laterally, and lastly posteriorly—and sink, converging ventrally under the embryo and cutting off the future gut cavity from the cavity of the yolk sac. The most anterior and posterior portions of the gut separate, but the middle part remains in open communication with the yolk sac throughout embryonic life, eventually becoming reduced to the yolk stalk, which passes through the umbilical cord.

The alimentary canal of vertebrates becomes differentiated into the oral cavity, pharynx, esophagus, stomach, and intestine. Whether derived from an archenteron or formed by folding of the endodermal sheet, the canal initially does not possess an opening at its anterior end. This is also the case in some lower chordates and echinoderms, which are grouped together with vertebrates as the Deuterostomia, or animals with secondary mouths.

In vertebrates, a mouth forms by a rupture at the anterior end, where the endoderm is in contact with ectoderm. The ectoderm of the future mouth region becomes depressed, forming a mouth invagination, or stomodaeum. The ectodermal and endodermal layers separating the cavity of the stomodaeum from the gut fuse to form the oropharyngeal membrane, which thins and ruptures, providing free passage from the exterior to the gut. Because of its mode of origin, the oral cavity is in part lined by ectoderm and in part by endoderm, the two parts becoming indistinguishable. Before the oropharyngeal membrane ruptures, however, a small pocket forms on the dorsal side of the stomodaeal invagination. This, the rudiment of the anterior lobe of the hypophysis, becomes apposed to the ventral surface of the diencephalon and loses its connection with the mouth cavity.

The anal opening in some exceptional cases (urodele amphibians) is derived directly from the blastopore, which persists as a narrow canal after completion of gastrulation. In other vertebrates, however, the anus develops either near the location of the former blastopore or in a corresponding region at the posterior end of the embryo, where the last remnants of mesoderm migrated to the interior. It is thus claimed that the anus in vertebrates is derived, directly or indirectly, from the blastopore. The mode of formation of the opening is somewhat similar to that of the mouth. A slight invagination of the ectoderm occurs, and a cloacal membrane forms, separating the ectodermal invagination from the gut cavity. The membrane ruptures later to provide the anus.

The Pharynx and its Outgrowths

The anterior portion of the endodermal gut, lying immediately posterior to the mouth cavity, expands laterally as the pharynx. The lateral pockets of the pharyngeal cavity, called the pharyngeal pouches, perforate the mesodermal layer, reach the ectoderm, and break through to form pharyngeal, or gill, clefts. In fishes and larvae of amphibians, these clefts develop gills and become respiratory organs. Pharyngeal pouches develop in the early embryos of all vertebrates, including the air-breathing terrestrial reptiles, birds, and mammals. The number of pouches has been reduced in the course of evolution from six or more to four in tetrapods, and the posterior pouches may not actually break through.

The consistent development of pharyngeal pouches and clefts indicates their importance in vertebrate development. Many parts of the vertebrate body are derived from, or dependent on, the pharyngeal pouches; for example, the aortic arches—the most important blood vessels of a vertebrate—develop between successive pharyngeal pouches. Skeletal visceral arches also occur between consecutive pharyngeal pouches (they do not develop if the pharyngeal pouches are prevented from developing). In adult terrestrial vertebrates, parts of the visceral arches are transformed into the hyoid apparatus, supporting the tongue, the auditory ossicles, and parts of the larynx and trachea. Furthermore, some of the material of the pharyngeal pouches is utilized for the formation of the parathyroid glands and the thymus; the former are indispensable glands of internal secretion, and the latter are a source, in mammals, of cells that produce antibodies. The pharynx also produces the rudiment of the thyroid gland as a ventral outgrowth.

The Liver, Pancreas and Lungs

Three additional important organs develop from the endoderm: the liver, the pancreas, and the lungs. The liver develops as a ventral outgrowth of the endodermal gut just posterior to the section that eventually will become the stomach. Initially, the liver takes the form of a tubular gland, but it soon acquires a close relationship to the blood sinuses and capillaries, forming lobules around blood vessels rather than around glandular ducts. The pancreas develops from three independent rudiments: two ventral ones, formed just posterior to the liver rudiment, and a dorsal one. The ventral and the dorsal rudiments fuse in most vertebrates to form one organ with a complicated system of ducts opening into the duodenum, a portion of the small intestine. The lungs develop from a ventral hollow outgrowth of the gut, which is located just posterior to the pharyngeal region; the outgrowth branches into a right and left trunk that grow posteriorly beside the esophagus and then expand into hollow sacs, in lower terrestrial vertebrates, or into a system of tubes, in birds and mammals.

The endodermal parts of the alimentary system are, along their entire length, encased by the splanchnic mesoderm of the lateral plates. The coelomic cavities of the right and left sides fuse ventral to the gut but remain separated dorsally by their respective walls, which form the dorsal mesentery—a double membrane by which the gut is suspended from the dorsal side of the body cavity and through which blood vessels and nerves reach the gut. The layer of splanchnic mesoderm next to the endoderm produces the connective tissue and muscular layers of the gut. During development of the glands of the alimentary canal (*e.g.*, pancreas, salivary glands), the mesoderm forms a connective tissue capsule around the branching tubules of the gland. The development of the tubules is dependent on this mesodermal capsule and cannot proceed without it.

Postembryonic Development

After partially developing within the egg membranes or within the maternal body, the newly formed individual emerges. The new animal is then born (ejected from the mother's body) or hatched from the egg. The condition of the new organism at the time of birth or hatching differs in various groups of animals, and even among animals within a particular group. In sea urchins, for example, the embryo emerges soon after fertilization, in the blastula stage. Covered with cilia, the sea-urchin blastula swims in the water and proceeds with gastrulation. Frog embryos emerge from the egg membranes when the main organs have already begun to develop, but functional differentiation of the tissues is unfinished; for instance, the components of the eyes and ears are far from complete, the mouth is not yet open, and the gut is filled with yolk-laden cells. Certain birds (called precocial) emerge from the egg covered with downy feathers and can run about soon after hatching, whereas others (altricial) hatch naked, with only rudiments of feathers, and are quite unable to move around. Among mammals there is a great range in the degree of development at birth. In marsupials, such as opossums and kangaroos, the young are born incompletely developed and very small; the young are then kept for a long time in the pouch of the mother, all the while firmly attached to the teats and suckling. Many small mammals are helpless at birth. Mice are born naked and blind; puppies and kittens are born covered with fur but with unopened eyes. Newborn human babies have their eyes open but cannot move themselves about for several months. Hoofed mammals, on the other hand, bear young that can stand up and run after their mothers within a few hours of birth.

In birds the hard shell is broken by the hatchling's beak, which is provided with a sharp tubercle on its top. A similar "egg tooth" appears on the tip of the snout of hatchling reptiles. Many arthropods have a preformed line of fragility that allows part of the eggshell to be burst open like a lid, allowing the young to emerge. Birth in mammals is effected through the contraction of smooth muscles of the uterus.

The Larval Phase and Metamorphosis

The organism emerging from the egg or from the maternal body, apart from being incompletely developed, may have an organization more or less different from that of an adult. In some cases the difference is so great that, without knowing the origin of the eggs or without following the young through their full course of development, it would be impossible to know that the young and the adult are of the same animal species. Such young, called larvae, transform into the adult form by a process of metamorphosis. The term larva also applies to young that resemble the adult form but differ from it in some substantial respect, as in possessing organs not present in the adult or in lacking an important structure (apart from sex glands and associated parts, which tend to develop later in life in most animals). Larvae in different animals have special names given to them, such as the tadpole of frogs, the caterpillar of butterflies, and the fry of fishes.

The Larval Stage

The development of the embryo into a larva rather than directly into an organism similar to the adult has various advantages. At the time of emergence from the egg, the new individual is relatively small, and the organization that enables the adult to lead a particular mode of life may not be suitable for a miniature copy of the adult. The larva may have to procure food for itself and, being small, may not be able to feed in the same way as the adult. It also may not be able to use effectively the same defense mechanisms the adult possesses. The larval stage enables an animal to avoid such hazards; it provides a mode of life and corresponding organization better suited to the smaller size of the newly emerged organism. Another advantage is that the larva may be able to exploit an entirely different environment because its organization is very different from that of the adults. A terrestrial adult may have aquatic larvae, a flying adult may have burrowing larvae, and a parasitic adult may have a free-living larva. A third advantage of a larval stage emerges in animals whose adult stages are sessile or restricted in their movements; the larvae can move freely, either of their own accord or on water currents. In this way the larvae of sedentary animals serve for the dispersal of the species. Lastly, the larval stage is of great advantage for certain internal parasites, which, once inside a host, cannot transfer to another. New hosts are infected instead by the larval stages. (The usual means of attaining this end is for the parasite to produce enormous quantities of eggs and rely on the passive entry of the eggs into the new host with food. A more efficient way, however, is for a mobile larva to enter the new host actively.)

A large number of marine invertebrates possess floating larvae that have hairlike projections (cilia) as their means of locomotion. There are three main types of larvae, characteristic of large subdivisions of the animal kingdom.

The planula larva of coelenterates has an elongated shape and cilia covering its entire surface. The internal organization is simple, hardly beyond differentiation into ectoderm and endoderm in the interior. The larva does not feed but serves only for dispersal.

The trochophore larva is found in many marine invertebrates. Typically, as in polychaetes, it has an alimentary canal with mouth and anus and a ring of ciliated cells arranged anterior to the mouth. It also possesses a sensory organ and rudiments of mesoderm. Cilia around the mouth bring in food—unicellular plants and other small particles. The larva thus not only serves for dispersal but also feeds and grows before it transforms into an adult worm. Other trochophore larvae are found in marine mollusks and in certain marine worms. The larva of echinoderms is similar to the trochophore in possessing a gut and a ciliary band, but the arrangement of the latter is different. The echinoderm larva also feeds and grows as well as serves for dispersal.

Larvae of very different kinds are found in many arthropods. In crustaceans the larva, called nauplius, does not differ substantially in mode of life or means of locomotion from the adult but has fewer appendages than the adult. A typical crustacean nauplius has three pairs of legs and an unpaired simple eye. Additional pairs of appendages and paired compound eyes appear in the course of a sometimes prolonged development. In insects the larva differs from the adult by the absence of wings but, in addition, may have a different mode of life and different way of feeding. Among chordates the tunicates (sea squirts) deserve attention; the larval form is a free-swimming creature, showing unmistakable relation to vertebrates, but the adult is sedentary, with much reduced nervous and muscular systems. The tadpole of a frog differs from the adult in being totally aquatic, in possessing a tail and gills for respiration, and in having a mouth adapted for feeding on plants. The adult frog is adapted to land life, except for reproductive periods, has no tail and no gills, and is an active predator.

Metamorphosis

Metamorphosis, the transformation of the larva into an adult, is a more or less complicated process depending on the degree of difference between the two forms. The transformation may be gradual, extend over a long period, and involve a number of intermediate stages; alternatively, the transformation may be achieved in one step. In the latter case, especially if the difference between the larva and adult is great, large parts of the body of the larva, including all the specifically larval organs, disintegrate (necrobiotic metamorphosis). At the same time, organs of the adult are built up, sometimes from reserve groups of cells that remain undifferentiated or nonfunctional in the larva. A good illustration of the distinction between gradual and abrupt metamorphosis occurs among the insects. In more primitive insects, such as cockroaches and grasshoppers, metamorphosis is gradual. The larva, often referred to as a nymph, has more or less the same organization as the adult, or imago; it feeds in a similar way but differs from the adults in lacking wings and in having incomplete sex organs. The wings appear in later stages of larval life; they are small at first but increase with each molt, and they attain full size and functional capacity at the last (imaginal) one. The larva of other insects, such as beetles, butterflies, and wasps, is a grub or caterpillar, a wormlike creature not even remotely resembling the adult. The difference in organization is so profound that the transformation cannot be achieved gradually, and an intermediate resting, or pupal, stage is interposed between the larva and imago. The pupa neither feeds nor moves, as the larval organs inside are destroyed and replaced with organs of the adult, including wings and sex organs. Eventually, when formation of the adult organs is complete, the pupal skin is cast off, and the adult emerges. The destruction of the larval parts may be far reaching and include even the skin and most of the alimentary canal. The tissues of the adult are formed from groups of reserve cells that were present all along in the larva as imaginal disks.

Necrobiotic metamorphosis is observed in the tunicate larva, in which the tail, including notochord, nerve cord, and muscles, and most of the brain, including eye and statocyst, are destroyed at the same time that the large pharyngeal cavity of the adult develops. A tadpole metamorphosing into an adult frog loses its tail—the cells of which are destroyed and devoured by phagocytic cells—its gills, and its larval mouthparts; concurrently the legs of the adult frog develop progressively, the structure of the mouth and alimentary canal change, and the skin acquires a bony (keratinized) layer and a system of subcutaneous glands.

The complicated changes taking place during metamorphosis, especially in the case of necrobiotic metamorphosis, must be performed in a coordinated way. So that no changes are made prematurely and no organ systems are left behind in the general transformation, some common signal for the change must be provided. For both insect and amphibian metamorphoses, which have been the most extensively studied, the signal is a hormonal one, sent in the blood to all the cells and tissues of the body.

Metamorphosis in an insect is complicated by the fact that the rigid cuticle covering its body is very restrictive; new features can appear only after a molt, when the old cuticle is replaced by a newly formed one. Molting in insects is caused by the action of two hormones. In the brain of insects, several groups of neurosecretory cells produce the first hormone. This brain hormone does not itself affect molting but stimulates the prothoracic gland, a loose mass of secretory cells situated in the thorax in close association with tracheal tubes. In response to the stimulation by the brain hormone, the prothoracic gland releases into the blood a second hormone, the molting hormone, or ecdysone. Under the influence of ecdysone, the tissues of the body produce a new cuticle under the old one, after which the old cuticle is shed (the actual molting). The new cuticle embodies any new developmental features that were scheduled to appear. The kind of feature that emerges after a molt is controlled by a third organ of internal secretion, the corpus allatum, secretory tissue situated posterior to the brain, near or around the dorsal aorta and usually appearing as a pair of separate or fused organs. The corpora allata emit the juvenile hormone, which, as long as it circulates in the blood, acts to perpetuate the larval form. As the larva approaches the end of its development, however, the corpora allata stop producing juvenile hormone or reduce its quantity; whereupon, the larva, at the next molt, metamorphoses into an adult. Withdrawal of the juvenile hormone is the immediate cause of metamorphosis, in conjunction with the brain hormone and ecdysone, which are responsible for the shedding of the larval cuticle and for the production of the new cuticle embodying the features of the imago. Metamorphosis through the stage of the pupa is effected by diminishing levels of juvenile hormone, which determine first the transformation of the larva into a pupa and, with further reduction of the juvenile-hormone level, the final step of transformation of the pupa into the adult.

The metamorphosis of a tadpole into a frog also depends upon two hormones: one initiating the process and the other directly influencing the tissues involved in the change. The first hormone is the thyrotropic hormone, produced by the hypophysis. It has no immediate effect on the tissues of the body but activates the thyroid gland to produce several substances, the most important of which is thyroxine. Thyroxine and other iodine-containing compounds circulate in the blood and cause changes that, in their entirety, constitute the process of metamorphosis. It is remarkable that different tissues react in different ways to the presence of thyroxine. The muscles of the tadpole's tail degenerate, whereas the muscles of the trunk and legs are not affected; in fact, the growth and development of limbs are stimulated as a part of metamorphosis. The effect of the hormone depends on the nature of the reacting cells and tissues—i.e., on their competence—just as the embryonic inductor in the earlier stages of development influences only cells with the competence for a particular kind of reaction.

Direct Development

If an animal after birth or emergence from an egg differs from the adult in comparatively minor details (apart from not having functional sex organs), the development is said to be direct. There is no larval stage and no metamorphosis. Direct development does not mean, however, that no changes occur between birth and adulthood. One very obvious change is the growth of the animal.

The rate of growth—not absolute increase—is highest in the early stages of postembryonic life; subsequently, growth continues to slow, ceasing completely at the attainment of adulthood. The rate of growth is dependent on many factors, both external (feeding, temperature) and internal. Of the internal factors, the most important are hormones, especially the growth hormone produced by the hypophysis. If the growth hormone is produced in insufficient quantities, the result is dwarfism; if it is produced in excessive quantities, the result is gigantism.

In the case of direct development, the most important change is the attainment of sexual maturity, which is achieved in several steps and involves the action of several hormones. The gonad rudiments and rudiments of the supporting parts of the reproductive system remain inactive long after birth. At the approach of adulthood, however, two sets of hormones come into action: hypophyseal hormones stimulate the gonads to activity, and gonadal hormones (produced by the gonads) cause the supporting sex organs and other sex characters to become fully developed. To become functional, the gonads must be acted upon by secretions from the hypophysis. In immature females the follicle-stimulating hormone, which alone causes the egg follicles and the oocytes to grow, and the luteinizing hormone stimulate the follicle cells to produce the female sex hormone, estrogen, which effects the development of the uterus, the milk glands, and other characteristics of the female sex. In the male, the same hypophyseal hormones are produced, with the result that the testes start to produce sperm and to secrete the male sex hormone, androgen. It appears that the luteinizing hormone is the more active in the male sex, being able to cause both spermatogenesis and androgen secretion. Androgen, in turn, brings about the development of the penis, the descent of the testicles before birth, the appearance of typical male hair growth, and other secondary sex characteristics.

Maturity and Death

Sexual maturity and the ensuing reproductive activity mark the pinnacle of development and morphogenesis and, for many animals, herald the end of life. The biological goal of the entire process is achieved with the launching of the next generation, and the life cycle that runs from the formation of gametes by one generation to the formation of gametes by the next generation is completed. In many animals the females die after laying their eggs; the males may have died earlier, after pairing. Indeed, some males (spiders, praying mantises) are eaten by the females immediately after copulation.

The developmental period can only truly be said to end with the termination of an organism, for much activity continues to unfold new developmental sequences, not all of them progressive and favourable, to be sure. Senescence, or a decline in abilities, signals advancing age in mammals but is not a general occurrence in the animal kingdom. Far more animals continue to function at near-peak capacity well into old age. And even among those species—salmons, eels, many moths—whose members die after a single reproductive act, death is relatively swift and not accompanied by a prolonged period of deterioration.

Mating

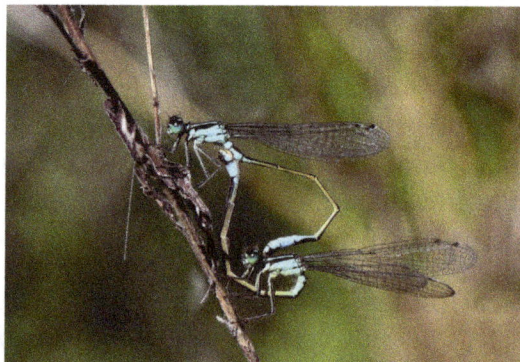

Blue-tailed damselflies (Ischnura elegans) mating.

Mating is the pairing of either opposite-sex or hermaphroditic organisms, usually for the purposes of sexual reproduction. Some definitions limit the term to pairing between animals, while other definitions extend the term to mating in plants and fungi. Fertilization is the fusion of both sex cell or gamete. Copulation is the union of the sex organs of two sexually reproducing animals for insemination and subsequent internal fertilization. Mating may also lead to external fertilization, as seen in amphibians, fishes and plants. For the majority of species, mating is between two individuals of opposite sexes. However, for some hermaphroditic species, copulation is not required because the parent organism is capable of self-fertilization (autogamy); for example, banana slugs.

The term mating is also applied to related processes in bacteria, archaea and viruses. Mating in these cases involves the pairing of individuals, accompanied by the pairing of their homologous chromosomes and then exchange of genomic information leading to formation of recombinant progeny.

Animals

For animals, mating strategies include random mating, disassortative mating, assortative mating, or a mating pool. In some birds, it includes behaviors such as nest-building and feeding offspring. The human practice of mating and artificially inseminating domesticated animals is part of animal husbandry.

In some terrestrial arthropods, including insects representing basal (primitive) phylogenetic clades, the male deposits spermatozoa on the substrate, sometimes stored within a special structure. Courtship involves inducing the female to take up the sperm package into her genital opening without actual copulation. In groups such as dragonflies and many spiders, males extrude sperm into secondary copulatory structures removed from their genital opening, which are then used to inseminate the female (in dragonflies, it is a set of modified sternites on the second abdominal segment; in spiders, it is the male pedipalps). In advanced groups of insects, the male uses its aedeagus, a structure formed from the terminal segments of the abdomen, to deposit sperm directly (though sometimes in a capsule called a "spermatophore") into the female's reproductive tract.

Other animals reproduce sexually with external fertilization, including many basal vertebrates. Vertebrates (such as reptiles, some fish, and most birds) reproduce with internal fertilization through cloacal copulation , while mammals copulate vaginally.

In domesticated animals there are various type of mating methods being employed to mate animals like Pen Mating (when female is moved to the desired male into a pen) or paddock mating (where one male is let loose in the paddock with several females).

Gray wolves mating.

Lions mating.

African spurred tortoises
(Centrochelys sulcata) mating.

Ladybugs mating.

Spittlebugs (Aphrophora alni) mating.

Plants and Fungi

Like in animals, mating in other Eukaryotes, such as plants and fungi, denotes sexual conjugation. However, in vascular plants this is mostly achieved without physical contact between mating individuals , and in some cases, e.g., in fungi no distinguishable male or female organs exist; however, mating types in some fungal species are somewhat analogous to sexual dimorphism in animals, and determine whether or not two individual isolates can mate. *Yeasts* are eukaryotic microorganisms classified in the kingdom Fungi, with 1,500 species currently described. In general, under high stress conditions like nutrient starvation, haploid cells will die; under the same conditions, however, diploid cells of *Saccharomyces cerevisiae* can undergo sporulation, entering sexual reproduction (meiosis) and produce a variety of haploid spores, which can go on to mate (conjugate) and reform the diploid.

Protists

Protists are a large group of diverse eukaryotic microorganisms, mainly unicellular animals and plants, that do not form tissues. Eukaryotes emerged in evolution more than 1.5 billion years ago. The earliest eukaryotes were likely protists. Mating and sexual reproduction are widespread among extant eukaryotes including protists such as Paramecium and Chlamydomonas. In many eukaryotic species, mating is promoted by sex pheromones including the protist Blepharisma japonicum. Based on a phylogenetic analysis, Dacks and Roger proposed that facultative sex was present in the common ancestor of all eukaryotes.

However, to many biologists it seemed unlikely until recently, that mating and sex could be a primordial and fundamental characteristic of eukaryotes. A principal reason for this view was that mating and sex appeared to be lacking in certain pathogenic protists whose ancestors branched off early from the eukaryotic family tree. However, several of these protists are now known to be capable of, or to recently have had, the capability for meiosis and hence mating. To cite one example, the common intestinal parasite Giardia intestinalis was once considered to be a descendant of a protist lineage that predated the emergence of meiosis and sex. However, G. intestinalis was recently found to have a core set of genes that function in meiosis and that are widely present among sexual eukaryotes. These results suggested that G. intestinalis is capable of meiosis and thus mating and sexual reproduction. Furthermore, direct evidence for meiotic recombination, indicative of mating and sexual reproduction, was also found in G. intestinalis. Other protists for which evidence of mating and sexual reproduction has recently been described are parasitic protozoa of the genus Leishmania, Trichomonas vaginalis, and acanthamoeba.

Protists generally reproduce asexually under favorable environmental conditions, but tend to reproduce sexually under stressful conditions, such as starvation or heat shock.

Fertilization

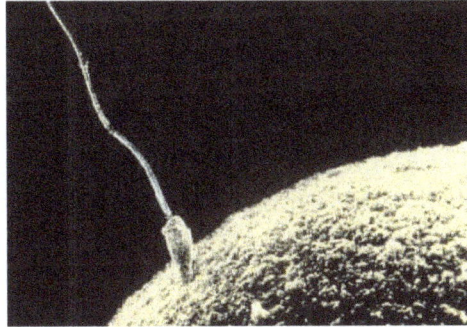

Sperm and ovum fusing.

Fertilisation or fertilization, also known as generative fertilisation, insemination, pollination, fecundation, syngamy and impregnation, is the fusion of gametes to initiate the development of a new individual organism or offspring. This cycle of fertilisation and development of new individuals is called sexual reproduction. During double fertilisation in angiosperms the haploid male gamete combines with two haploid polar nuclei to form a triploid primary endosperm nucleus by the process of vegetative fertilisation.

Evolution

The evolution of fertilisation is related to the origin of meiosis, as both are part of sexual reproduction, originated in eukaryotes. There are two conflicting theories on how the couple meiosis–fertilisation arose. One is that it evolved from prokaryotic sex (bacterial recombination) as eukaryotes evolved from prokaryotes. The other is that mitosis originated meiosis.

Fertilisation in Animals

The mechanics behind fertilisation has been studied extensively in sea urchins and mice. This research addresses the question of how the sperm and the appropriate egg find each other and the question of how only one sperm gets into the egg and delivers its contents. There are three steps to fertilisation that ensure species-specificity:

- Chemotaxis.
- Sperm activation/acrosomal reaction.
- Sperm/egg adhesion.

Internal vs. External

Consideration as to whether an animal (more specifically a vertebrate) uses internal or external fertilisation is often dependent on the method of birth. Oviparous animals laying eggs with thick

100 | Introduction to Developmental Biology

calcium shells, such as chickens, or thick leathery shells generally reproduce via internal fertilisation so that the sperm fertilises the egg without having to pass through the thick, protective, tertiary layer of the egg. Ovoviviparous and viviparous animals also use internal fertilisation. It is important to note that although some organisms reproduce via amplexus, they may still use internal fertilisation, as with some salamanders. Advantages to internal fertilisation include: minimal waste of gametes; greater chance of individual egg fertilisation, relatively "longer" time period of egg protection, and selective fertilisation; many females have the ability to store sperm for extended periods of time and can fertilise their eggs at their own desire.

Oviparous animals producing eggs with thin tertiary membranes or no membranes at all, on the other hand, use external fertilisation methods. Advantages to external fertilisation include: minimal contact and transmission of bodily fluids; decreasing the risk of disease transmission, and greater genetic variation (especially during broadcast spawning external fertilisation methods).

Sea Urchins

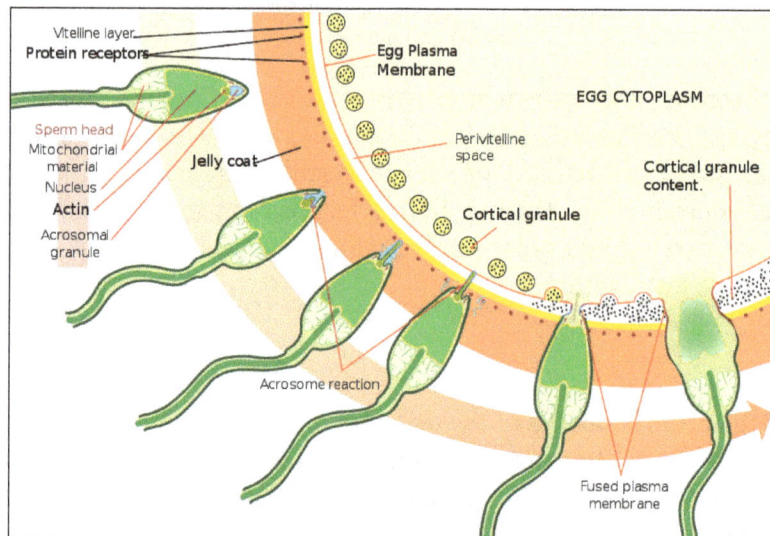

Acrosome reaction on a sea urchin cell.

Sperm find the eggs via chemotaxis, a type of ligand/receptor interaction. Resact is a 14 amino acid peptide purified from the jelly coat of A. punctulata that attracts the migration of sperm.

After finding the egg, the sperm penetrates the jelly coat through a process called sperm activation. In another ligand/receptor interaction, an oligosaccharide component of the egg binds and activates a receptor on the sperm and causes the acrosomal reaction. The acrosomal vesicles of the sperm fuse with the plasma membrane and are released. In this process, molecules bound to the acrosomal vesicle membrane, such as bindin, are exposed on the surface of the sperm. These contents digest the jelly coat and eventually the vitelline membrane. In addition to the release of acrosomal vesicles, there is explosive polymerisation of actin to form a thin spike at the head of the sperm called the acrosomal process.

The sperm binds to the egg through another ligand reaction between receptors on the vitelline membrane. The sperm surface protein bindin, binds to a receptor on the vitelline membrane identified as EBR1.

Fusion of the plasma membranes of the sperm and egg are likely mediated by bindin. At the site of contact, fusion causes the formation of a fertilisation cone.

Mammals

Mammals internally fertilise through copulation. After a male ejaculates, many sperm move to the upper vagina (via contractions from the vagina) through the cervix and across the length of the uterus to meet the ovum. In cases where fertilisation occurs, the female usually ovulates during a period that extends from hours before copulation to a few days after; therefore, in most mammals it is more common for ejaculation to precede ovulation than vice versa.

When sperm cells are deposited into the anterior vagina, they are not capable of fertilization (i.e., non-capacitated) and are characterized by slow linear motility patters. This motility pattern, combined with muscular contractions enables sperm transport towards the uterus and fallopian tubes. There is a pH gradient within the microenvironment of the female reproductive tract such that the pH near the vaginal opening is lower (approximately 5.0) than the fallopian tubes (approximately 8.0). The sperm-specific pH-sensitive calcium transport protein called CatSper increases the sperm cell permeability to calcium as it moves further into the reproductive tract. Intracellular calcium influx contributes to sperm capacitation and hyperactivation, causing a more violent and rapid non-linear motility pattern as the sperm approach the oocyte. The capacitated spermatozoon and the oocyte meet and interact in the *ampulla* of the fallopian tube. Rheotaxis, thermotaixs and chemotaxis are known mechanisms in guiding sperm towards the egg during the final stage of sperm migration. Spermatozoa respond to the temperature gradient of ~2 °C between the oviduct and the ampulla, and chemotactic gradients of progesterone have been confirmed as the signal emanating from the cumulus oophorus cells surrounding rabbit and human oocytes. Capacitated and hyperactivated sperm respond to these gradients by changing their behaviour and moving towards the cumulus-oocyte complex. Other chemotactic signals such as formyl Met-Leu-Phe (fMLF) may also guide spermatozoa.

The zona pellucida, a thick layer of extracellular matrix that surrounds the egg and is similar to the role of the vitelline membrane in sea urchins, binds with the sperm. Unlike sea urchins, the sperm binds to the egg before the acrosomal reaction. ZP3, a glycoprotein in the zona pellucida, is responsible for egg/sperm adhesion in mice. The receptor galactosyltransferase (GalT) binds to the N-acetylglucosamine residues on the ZP3 and is important for binding with the sperm and activating the acrosome reaction. ZP3 is sufficient though unnecessary for sperm/egg binding. Two additional sperm receptors exist: a 250kD protein that binds to an oviduct secreted protein, and SED1, which independently binds to the zona. After the acrosome reaction, the sperm is believed to remain bound to the zona pellucida through exposed ZP2 receptors. These receptors are unknown in mice but have been identified in guinea pigs.

In mammals, the binding of the spermatozoon to the GalT initiates the acrosome reaction. This process releases the hyaluronidase that digests the matrix of hyaluronic acid in the vestments around the oocyte. Additionally, heparin-like glycosaminoglycans (GAGs) are released near the oocyte that promote the acrosome reaction. Fusion between the oocyte plasma membranes and sperm follows and allows the sperm nucleus, the typical centriole, and atypical centriole that is attached to the flagellum, but not the mitochondria, to enter the oocyte. The protein CD9 likely mediates this fusion in mice (the binding homolog). The egg "activates" itself upon fusing with a single sperm cell and thereby changes its cell membrane to prevent fusion with other sperm. Zinc atoms are released during this activation.

This process ultimately leads to the formation of a diploid cell called a zygote. The zygote divides to form a blastocyst and, upon entering the uterus, implants in the endometrium, beginning pregnancy. Embryonic implantation not in the uterine wall results in an ectopic pregnancy that can kill the mother.

In such animals as rabbits, coitus induces ovulation by stimulating the release of the pituitary hormone gonadotropin; this release greatly increases the likelihood of pregnancy.

Humans

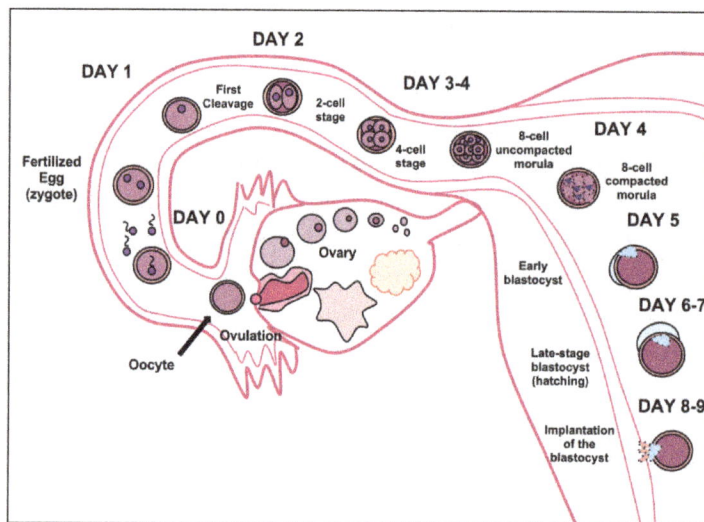

The process of fertilisation in humans.

Fertilisation in humans is the union of a human egg and sperm, usually occurring in the ampulla of the fallopian tube, producing a zygote cell, or fertilized egg, initiating prenatal development. Scientists discovered the dynamics of human fertilization in the nineteenth century.

The term *conception* commonly refers to "the process of becoming pregnant involving fertilization or implantation or both". Its use makes it a subject of semantic arguments about the beginning of pregnancy, typically in the context of the abortion debate. Upon gastrulation, which occurs around 16 days after fertilisation, the implanted blastocyst develops three germ layers, the endoderm, the ectoderm and the mesoderm, and the genetic code of the father becomes fully involved in the development of the embryo; later twinning is impossible. Additionally, interspecies hybrids survive only until gastrulation and cannot further develop. However, some human developmental biology literature refers to the *conceptus* and such medical literature refers to the "products of conception" as the post-implantation embryo and its surrounding membranes. The term "conception" is not usually used in scientific literature because of its variable definition and connotation.

Insects

Insects in different groups, including the Odonata (dragonflies and damselflies) and the Hymenoptera (ants, bees, and wasps) practise delayed fertilisation. Among the Odonata, females may mate with multiple males, and store sperm until the eggs are laid. The male may hover above the female during egg-laying (oviposition) to prevent her from mating with other males and replacing his sperm; in some groups such as the darters, the male continues to grasp the female with his claspers during egg-laying, the pair flying around in tandem. Among social Hymenoptera, honeybee queens mate only on mating flights, in a short period lasting some days; a queen may mate with eight or more drones. She then stores the sperm for the rest of her life, perhaps for five years or more.

Red-veined darters (Sympetrum fonscolombii) flying "in cop" (male ahead), enabling the male to prevent other males from mating. The eggs are fertilised as they are laid, one at a time.

Fertilisation in Fungi

In many fungi (except chytrids), as in some protists, fertilisation is a two step process. First, the cytoplasms of the two gamete cells fuse (called plasmogamy), producing a dikaryotic or heterokaryotic cell with multiple nuclei. This cell may then divide to produce dikaryotic or heterokaryotic hyphae. The second step of fertilisation is karyogamy, the fusion of the nuclei to form a diploid zygote.

In chytrid fungi, fertilisation occurs in a single step with the fusion of gametes, as in animals and plants.

Fertilisation in Protists

Fertilisation in Protozoa

There are three types of fertilisation processes in protozoa:

- Gametogamy;

- Autogamy;

- Gamontogamy.

Fertilisation in Algae

Fertilisation in algae occurs by binary fission. The pseudopodia is first withdrawn and the nucleus starts dividing. When the cytoplasm is divided, the cytoplasm is also divided into two equal parts for each daughter cell. Two daughter cells are produced by one parent cell. It involves the process of mitosis.

Fertilisation in Fungi-like Protists

Fertilisation in fungi. In many fungi (except chytrids), as in some protists, fertilisation is a two step process. In chytrid fungi, fertilisation occurs in a single step with the fusion of gametes, as in animals and plants.

Fertilisation and Genetic Recombination

Meiosis results in a random segregation of the genes that each parent contributes. Each parent organism is usually identical save for a fraction of their genes; each gamete is therefore genetically unique. At fertilisation, parental chromosomes combine. In humans, $(2^{22})^2 = 17.6 \times 10^{12}$ chromosomally different zygotes are possible for the non-sex chromosomes, even assuming no chromosomal crossover. If crossover occurs once, then on average $(4^{22})^2 = 309 \times 10^{24}$ genetically different zygotes are possible for every couple, not considering that crossover events can take place at most points along each chromosome. The X and Y chromosomes undergo no crossover events and are therefore excluded from the calculation. The mitochondrial DNA is only inherited from the maternal parent.

Parthenogenesis

Organisms that normally reproduce sexually can also reproduce via parthenogenesis, wherein an unfertilised female gamete produces viable offspring. These offspring may be clones of the mother, or in some cases genetically differ from her but inherit only part of her DNA. Parthenogenesis occurs in many plants and animals and may be induced in others through a chemical or electrical stimulus to the egg cell. In 2004, Japanese researchers led by Tomohiro Kono succeeded after 457 attempts to merge the ova of two mice by blocking certain proteins that would normally prevent the possibility; the resulting embryo normally developed into a mouse.

Allogamy and Autogamy

Allogamy, which is also known as cross-fertilisation, refers to the fertilisation of an egg cell from one individual with the male gamete of another.

Autogamy which is also known as self-fertilisation, occurs in such hermaphroditic organisms as plants and flatworms; therein, two gametes from one individual fuse.

Cleavage

Cleavage is the division of cells in the early embryo. The process follows fertilization, with the transfer being triggered by the activation of a cyclin-dependent kinase complex. The zygotes of many species undergo rapid cell cycles with no significant overall growth, producing a cluster of cells the same size as the original zygote. The different cells derived from cleavage are called blastomeres and form a compact mass called the morula. Cleavage ends with the formation of the blastula.

Depending mostly on the amount of yolk in the egg, the cleavage can be holoblastic (total or entire cleavage) or meroblastic (partial cleavage). The pole of the egg with the highest concentration of yolk is referred to as the vegetal pole while the opposite is referred to as the animal pole.

Cleavage differs from other forms of cell division in that it increases the number of cells and nuclear mass without increasing the cytoplasmic mass. This means that with each successive subdivision, there is roughly half the cytoplasm in each daughter cell than before that division, and thus the ratio of nuclear to cytoplasmic material increases.

Mechanism

The rapid cell cycles are facilitated by maintaining high levels of proteins that control cell cycle progression such as the cyclins and their associated cyclin-dependent kinases (cdk). The complex Cyclin B/CDK1 a.k.a. MPF (maturation promoting factor) promotes entry into mitosis.

The processes of karyokinesis (mitosis) and cytokinesis work together to result in cleavage. The mitotic apparatus is made up of a central spindle and polar asters made up of polymers of tubulin protein called microtubules. The asters are nucleated by centrosomes and the centrosomes are organized by centrioles brought into the egg by the sperm as basal bodies. Cytokinesis is mediated by the contractile ring made up of polymers of actin protein called microfilaments. Karyokinesis and cytokinesis are independent but spatially and temporally coordinated processes. While mitosis can occur in the absence of cytokinesis, cytokinesis requires the mitotic apparatus.

The end of cleavage coincides with the beginning of zygotic transcription. This point is referred to as the midblastula transition and appears to be controlled by the nuclear:cytoplasmic ratio (about 1/6).

Types of Cleavage

Determinate

Determinate cleavage (also called mosaic cleavage) is in most protostomes. It results in the developmental fate of the cells being set early in the embryo development. Each blastomere produced by early embryonic cleavage does not have the capacity to develop into a complete embryo.

Indeterminate

A cell can only be indeterminate (also called regulative) if it has a complete set of undisturbed animal/vegetal cytoarchitectural features. It is characteristic of deuterostomes – when the original cell in a deuterostome embryo divides, the two resulting cells can be separated, and each one can individually develop into a whole organism.

Holoblastic

In the absence of a large concentration of yolk, four major cleavage types can be observed in isolecithal cells (cells with a small even distribution of yolk) or in mesolecithal cells or microlecithal cells (moderate amount of yolk in a gradient) – bilateral holoblastic, radial holoblastic, rotational holoblastic, and spiral holoblastic, cleavage. These holoblastic cleavage planes pass all the way through isolecithal zygotes during the process of cytokinesis. Coeloblastula is the next stage of development for eggs that undergo these radial cleavaging. In holoblastic eggs, the first cleavage always occurs along the vegetal-animal axis of the egg, the second cleavage is perpendicular to the first. From here, the spatial arrangement of blastomeres can follow various patterns, due to different planes of cleavage, in various organisms.

- Bilateral:

  The first cleavage results in bisection of the zygote into left and right halves. The following cleavage planes are centered on this axis and result in the two halves being mirror images of one another. In bilateral holoblastic cleavage, the divisions of the blastomeres are complete and separate; compared with bilateral meroblastic cleavage, in which the blastomeres stay partially connected.

- Radial:

  Radial cleavage is characteristic of the deuterostomes, which include some vertebrates and echinoderms, in which the spindle axes are parallel or at right angles to the polar axis of the oocyte.

- Rotational:

  Rotational cleavage involves a normal first division along the meridional axis, giving rise to two daughter cells. The way in which this cleavage differs is that one of the daughter cells divides meridionally, whilst the other divides equatorially.

  Mammals display rotational cleavage, and an isolecithal distribution of yolk (sparsely and evenly distributed). Because the cells have only a small amount of yolk, they require immediate implantation onto the uterine wall in order to receive nutrients.

  The nematode *C. elegans*, a popular developmental model organism, undergoes holoblastic rotational cell cleavage.

- Spiral:

  Spiral cleavage is conserved between many members of the lophotrochozoan taxa, referred to as Spiralia. Most spiralians undergo equal spiral cleavage, although some undergo unequal cleavage . This group includes annelids, molluscs, and sipuncula. Spiral cleavage can vary between species, but generally the first two cell divisions result in four macromeres, also called blastomeres, (A, B, C, D) each representing one quadrant of the embryo. These first two cleavages are not oriented in planes that occur at right angles parallel to the animal-vegetal axis of the zygote. At the 4-cell stage, the A and C macromeres meet at the animal pole, creating the animal cross-furrow, while the B and D macromeres meet at the vegetal pole, creating the vegetal cross-furrow. With each successive cleavage cycle, the macromeres give rise to quartets of smaller micromeres at the animal pole. The divisions that produce these quartets occur at an oblique angle, an angle that is not a multiple of 90°, to the animal-vegetal axis. Each quartet of micromeres is rotated relative to their parent macromere, and the chirality of this rotation differs between odd and even numbered quartets, meaning that there is alternating symmetry between the odd and even quartets. In other words, the orientation of divisions that produces each quartet alternates between being clockwise and counterclockwise with respect to the animal pole. The alternating cleavage pattern that occurs as the quartets are generated produces quartets of micromeres that reside in the cleavage furrows of the four macromeres. When viewed from the animal pole, this arrangement of cells displays a spiral pattern.

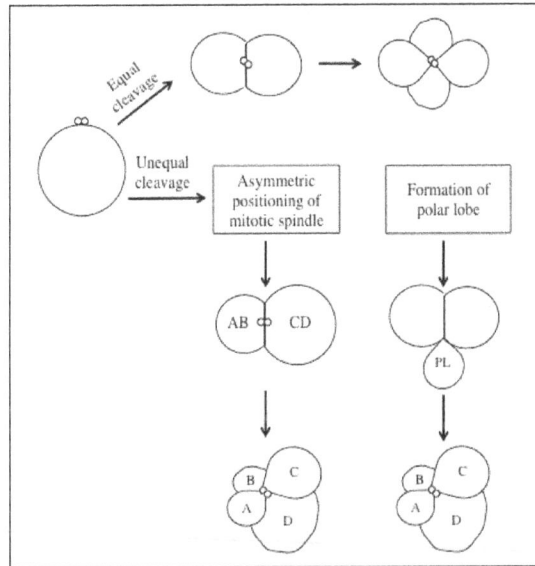

D quadrant specification through equal and unequal cleavage mechanisms. At the 4-cell stage of equal cleavage, the D macromere has not been specified yet. It will be specified after the formation of the third quartet of micromeres. Unequal cleavage occurs in two ways: asymmetric positioning of the mitotic spindle, or through the formation of a polar lobe (PL).

Specification of the D macromere and is an important aspect of spiralian development. Although the primary axis, animal-vegetal, is determined during oogenesis, the secondary axis, dorsal-ventral, is determined by the specification of the D quadrant. The D macromere facilitates cell divisions that differ from those produced by the other three macromeres. Cells of the D quadrant give rise to dorsal and posterior structures of the spiralian. Two known mechanisms exist to specify the D quadrant. These mechanisms include equal cleavage and unequal cleavage.

In equal cleavage, the first two cell divisions produce four macromeres that are indistinguishable from one another. Each macromere has the potential of becoming the D macromere. After the formation of the third quartet, one of the macromeres initiates maximum contact with the overlying micromeres in the animal pole of the embryo. This contact is required to distinguish one macromere as the official D quadrant blastomere. In equally cleaving spiral embryos, the D quadrant is not specified until after the formation of the third quartet, when contact with the micromeres dictates one cell to become the future D blastomere. Once specified, the D blastomere signals to surrounding micromeres to lay out their cell fates.

In unequal cleavage, the first two cell divisions are unequal producing four cells in which one cell is bigger than the other three. This larger cell is specified as the D macromere. Unlike equally cleaving spiralians, the D macromere is specified at the four-cell stage during unequal cleavage. Unequal cleavage can occur in two ways. One method involves asymmetric positioning of the cleavage spindle. This occurs when the aster at one pole attaches to the cell membrane, causing it to be much smaller than the aster at the other pole. This results in an unequal cytokinesis, in which both macromeres inherit part of the animal region of the egg, but only the bigger macromere inherits the vegetal region. The second mechanism of unequal cleavage involves the production of an enucleate, membrane bound, cytoplasmic protrusion, called a polar lobe. This polar lobe forms at the vegetal pole during cleavage, and then gets shunted to the D blastomere. The polar lobe contains vegetal cytoplasm, which becomes inherited by the future D macromere.

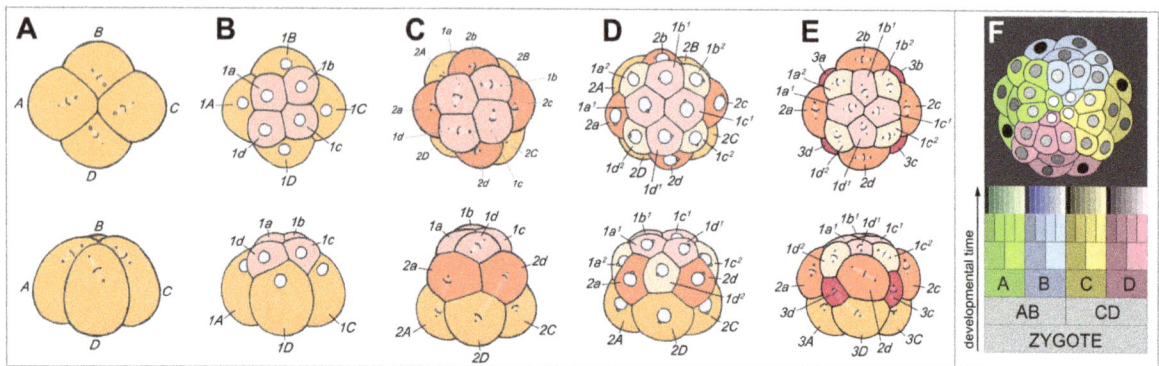

Spiral cleavage in marine snail of the genus Trochus.

Meroblastic

In the presence of a large amount of yolk in the fertilized egg cell, the cell can undergo partial, or meroblastic, cleavage. Two major types of meroblastic cleavage are discoidal and superficial:

- Discoidal

  In discoidal cleavage, the cleavage furrows do not penetrate the yolk. The embryo forms a disc of cells, called a blastodisc, on top of the yolk. Discoidal cleavage is commonly found in monotremes, birds, reptiles, and fish that have telolecithal egg cells (egg cells with the yolk concentrated at one end). The layer of cells that have incompletely divided and are in contact with the yolk are called the "syncytial layer".

- Superficial

  In superficial cleavage, mitosis occurs but not cytokinesis, resulting in a polynuclear cell. With the yolk positioned in the center of the egg cell, the nuclei migrate to the periphery of the egg, and the plasma membrane grows inward, partitioning the nuclei into individual cells. Superficial cleavage occurs in arthropods that have centrolecithal egg cells (egg cells with the yolk located in the center of the cell). This type of cleavage can work to promote synchronicity in developmental timing, such as in Drosophila.

Placentals

Differences exist between the cleavage in placental mammals and the cleavage in other animals.

Mammals have a slow rate of division that is between 12 and 24 hours. These cellular divisions are asynchronous. Zygotic transcription starts at the two-, four-, or eight-cell stage. Cleavage is holoblastic and rotational.

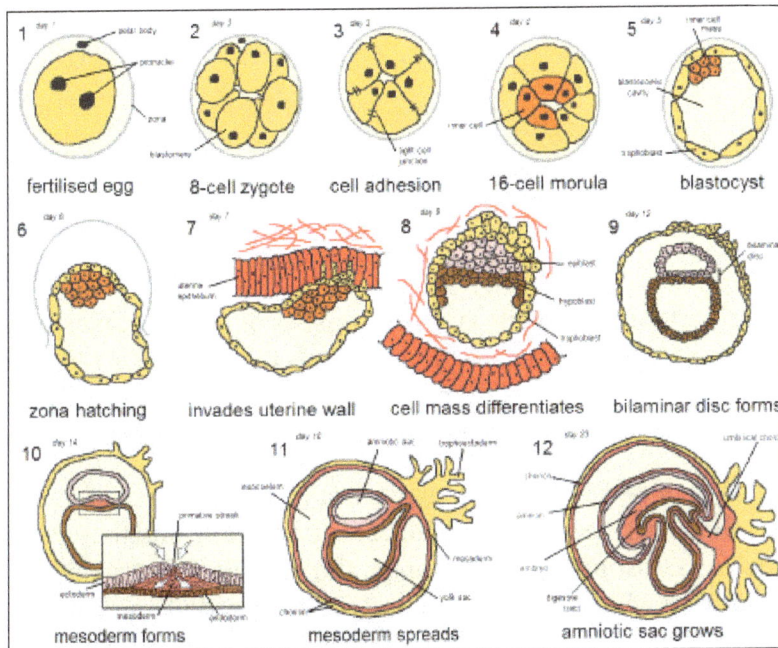

Human embryogenesis.

At the eight-cell stage, having undergone three cleavages the embryo goes through some changes. At this stage the cells begin to tightly adhere in a process known as compaction. Recently, it has been proposed that in placental mammals the cells become more likely to contribute to one of the first two cell types to arise, the inner cell mass or trophectoderm, depending on their position within the compacted embryo. A single cell can be removed from a pre-compaction eight- cell embryo and used for genetic testing and the embryo will recover.

Most of the blastomeres in this stage become polarized and develop tight junctions with the other blastomeres. This process leads to the development of two different populations of cells: Polar cells on the outside and apolar cells on the inside. The outer cells, called the trophoblast cells, pump sodium in from the outside, which automatically brings water in with it to the basal (inner) surface to form a blastocoel cavity in a process called cavitation. The trophoblast cells will eventually give rise to the embryonic contribution to the placenta called the chorion. The inner cells are pushed to one side of the cavity (because the embryo isn't getting any bigger) to form the inner cell mass (ICM) and will give rise to the embryo and some extraembryonic membranes. At this stage, the embryo is called a blastocyst.

Blastulation

The blastula is a hollow sphere of cells, referred to as blastomeres, surrounding an inner fluid-filled cavity called the blastocoele formed during an early stage of embryonic development in animals. Embryo development begins with a sperm fertilizing an egg to become a zygote which undergoes many cleavages to develop into a ball of cells called a morula. Only when the blastocoele is formed does the early embryo become a blastula. The blastula precedes the formation of the gastrula in which the germ layers of the embryo form.

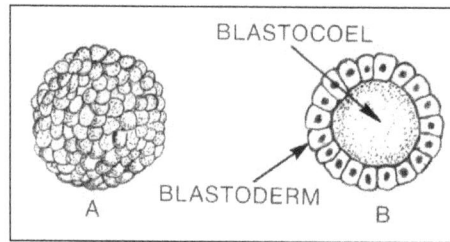

Blastocoel and blastoderm.

A common feature of a vertebrate blastula is that it consists of a layer of blastomeres, known as the blastoderm, which surrounds the blastocoele. In mammals the blastula is referred to as a blastocyst. The blastocyst contains an embryoblast (or inner cell mass) that will eventually give rise to the definitive structures of the fetus, and the trophoblast, which goes on to form the extra-embryonic tissues.

During the blastula stage of development, a significant amount of activity occurs within the early embryo to establish cell polarity, cell specification, axis formation, and to regulate gene expression. In many animals such as *Drosophila* and *Xenopus*, the mid blastula transition (MBT) is a crucial step in development during which the maternal mRNA is degraded and control over development is passed to the embryo. Many of the interactions between blastomeres are dependent on cadherin expression, particularly E-cadherin in mammals and EP-cadherin in amphibians.

The study of the blastula and of cell specification has many implications on the field of stem cell research as well as the continued improvement of fertility treatments. Embryonic stem cells are a field which, though controversial, have tremendous potential for treating disease. In *Xenopus*, blastomeres behave as pluripotent stem cells which can migrate down several pathways, depending on cell signaling. By manipulating the cell signals during the blastula stage of development, various tissues can be formed. This potential can be instrumental in regenerative medicine for disease and injury cases. In vitro fertilisation involves implantation of a blastula into a mother's uterus. Blastula cell implantation could serve to eliminate infertility.

Development

The blastula stage of early embryo development begins with the appearance of the blastocoel. The origin of the blastocoele in *Xenopus* has been shown to be from the first cleavage furrow, which is widened and sealed with tight junctions to create a cavity.

In many organisms the development of the embryo up to this point and for the early part of the blastula stage is controlled by maternal mRNA, so called because it was produced in the egg prior to fertilization and is therefore exclusively from the mother.

Mid-blastula Transition

In many organisms including Xenopus and Drosophila, the mid-blastula transition usually occurs after a particular number of cell divisions for a given species, and is defined by the ending of the synchronous cell division cycles of the early blastula development, and the lengthening of the cell cycles by the addition of the G1 and G2 phases. Prior to this transition, cleavage occurs with only the synthesis and mitosis phases of the cell cycle. The addition of the two growth phases into the cell cycle allows for the cells to increase in size, as up to this point the blastomeres undergo reductive divisions in which the overall size of the embryo does not increase, but more cells are created. This transition begins the growth in size of the organism.

The mid-blastula transition is also characterized by a marked increase in transcription of new, non-maternal mRNA transcribed from the genome of the organism. Large amounts of the maternal mRNA are destroyed at this point, either by proteins such as SMAUG in *Drosophila* or by microRNA. These two processes shift the control of the embryo from the maternal mRNA to the nuclei.

Structure

A blastula is a sphere of cells surrounding a blastocoele. The blastocoele is a fluid filled cavity which contains amino acids, proteins, growth factors, sugars, ions and other components which are necessary for cellular differentiation. The blastocoele also allows blastomeres to move during the process of gastrulation.

In Xenopus embryos, the blastula is composed of three different regions. The animal cap forms the roof of the blastocoele and goes on primarily to form ectodermal derivatives. The equatorial or marginal zone, which compose the walls of the blastocoel differentiate primarily into mesodermal tissue. The vegetal mass is composed of the blastocoel floor and primarily develops into endodermal tissue.

In the mammalian blastocyst (term for mammalian blastula) there are three lineages that give rise to later tissue development. The epiblast gives rise to the fetus itself while the trophoblast develops into part of the placenta and the primitive endoderm becomes the yolk sac.

In mouse embryo, blastocoele formation begins at the 32-cell stage. During this process, water enters the embryo, aided by an osmotic gradient which is the result of Na^+/K^+ ATPases that produce a high Na^+ gradient on the basolateral side of the trophectoderm. This movement of water is facilitated by aquaporins. A seal is created by tight junctions of the epithelial cells that line the blastocoele.

Cellular Adhesion

Tight junctions are very important in embryo development. In the blastula, these cadherin mediated cell interactions are essential to development of epithelium which are most important to paracellular transport, maintenance of cell polarity and the creation of a permeability seal to regulate blastocoel formation. These tight junctions arise after the polarity of epithelial cells is established which sets the foundation for further development and specification. Within the blastula, inner blastomeres are generally non-polar while epithelial cells demonstrate polarity.

Mammalian embryos undergo compaction around the 8-cell stage where E-cadherins as well as alpha and beta catenins are expressed. This process makes a ball of embryonic cells which are capable of interacting, rather than a group of diffuse and undifferentiated cells. E-cadherin adhesion defines the apico-basal axis in the developing embryo and turns the embryo from an indistinct ball of to a more polarized phenotype which sets the stage for further development into a fully formed blastocyst.

Xenopus membrane polarity is established with the first cell cleavage. Amphibian EP-cadherin and XB/U cadherin perform a similar role as E-cadherin in mammals establishing blastomere polarity and solidifying cell-cell interactions which are crucial for further development.

Clinical Implications

Fertilization Technologies

Experiments with implantation in mice show that hormonal induction, superovulation and artificial insemination successfully produce preimplantion mice embryos. In the mice, ninety percent of the females were induced by mechanical stimulation to undergo pregnancy and implant at least one embryo. These results prove to be encouraging because they provide a basis for potential implantation in other mammalian species, such as humans.

Stem Cells

Blastula-stage cells can behave as pluripotent stem cells in many species. Pluripotent stem cells are the starting point to produce organ specific cells that can potentially aid in repair and prevention of injury and degeneration. Combining the expression of transcription factors and locational positioning of the blastula cells can lead to the development of induced functional organs and tissues. Pluripotent *Xenopus* cells, when used in an in vivo strategy, were able to form into functional retinas. By transplanting them to the eye field on the neural plate, and by inducing several mis-expressions of transcription factors, the cells were committed to the retinal lineage and could guide vision based behavior in the *Xenopus*.

Implantation

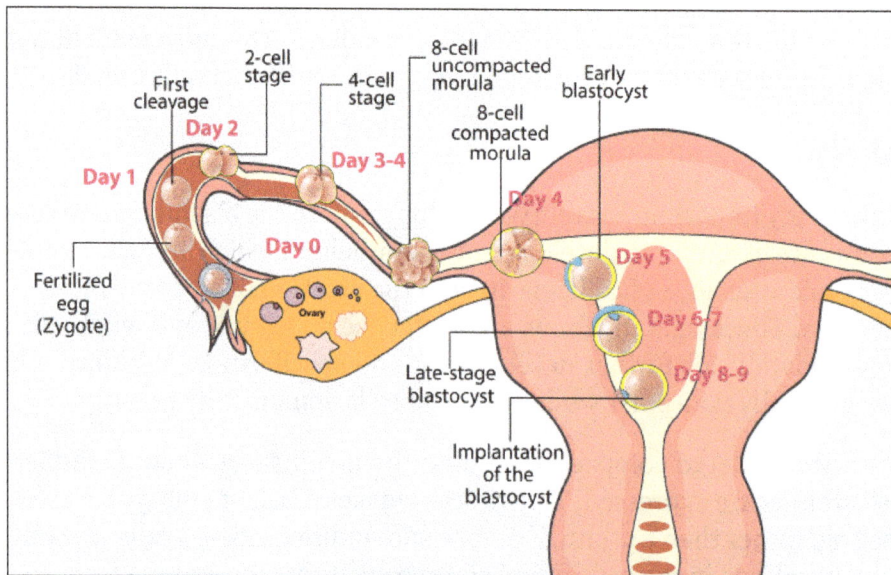

The sperm and ovum unite through fertilization, creating a conceptus that
(over the course of 8-9 days) will implant in the uterine wall, where
it will reside over the course of nine months.

Implantation is the stage of pregnancy at which the embryo adheres to the wall of the uterus. At this stage of prenatal development, the conceptus is called a blastocyst. It is by this adhesion that the embryo receives oxygen and nutrients from the mother to be able to grow.

In humans, implantation of a fertilized ovum is most likely to occur around nine days after ovulation; however, this can range between six and 12 days.

Implantation Window

The reception-ready phase of the endometrium of the uterus is usually termed the "implantation window" and lasts about 4 days. The implantation window occurs around 6 days after the peak in luteinizing hormone levels. With some disparity between sources, it has been stated to occur from 7 days after ovulation until 9 days after ovulation, or days 6-10 postovulation. On average, it occurs during the 20th to the 23rd day after the last menstrual period.

The implantation window is characterized by changes to the endometrium cells, which aid in the absorption of the uterine fluid. These changes are collectively known as the plasma membrane transformation and bring the blastocyst nearer to the endometrium and immobilize it. During this stage the blastocyst can still be eliminated by being flushed out of the uterus. Scientists have hypothesized that the hormones cause a swelling that fills the flattened out uterine cavity just prior to this stage, which may also help press the blastocyst against the endometrium. The implantation window may also be initiated by other preparations in the endometrium of the uterus, both structurally and in the composition of its secretions.

Adaptation of Uterus

To enable implantation, the uterus goes through changes in order to be able to receive the conceptus.

Predecidualization

The endometrium increases thickness, becomes vascularized and its glands grow to be tortuous and boosted in their secretions. These changes reach their maximum about 7 days after ovulation. Furthermore, the surface of the endometrium produces a kind of rounded cells, which cover the whole area toward the uterine cavity. This happens about 9 to 10 days after ovulation. These cells are called decidual cells, which emphasises that the whole layer of them is shed off in every menstruation if no pregnancy occurs, just as leaves of deciduous trees. The uterine glands, on the other hand, decrease in activity and degenerate around 8 to 9 days after ovulation in absence of pregnancy.

The decidual cells originate from the stromal cells that are always present in the endometrium. However, the decidual cells make up a new layer, the decidua. The rest of the endometrium, in addition, expresses differences between the luminal and the basal sides. The luminal cells form the zona compacta of the endometrium, in contrast to the basalolateral zona spongiosa, which consists of the rather spongy stromal cells.

Decidualization

Decidualization succeeds predecidualization if pregnancy occurs. This is an expansion of it, further developing the uterine glands, the zona compacta and the epithelium of decidual cells lining it. The decidual cells become filled with lipids and glycogen and take the polyhedral shape characteristic for decidual cells.

Trigger

It is likely that the blastocyst itself makes the main contribution to this additional growing and sustaining of the decidua. An indication of this is that decidualization occurs at a higher degree in conception cycles than in nonconception cycles. Furthermore, similar changes are observed when giving stimuli mimicking the natural invasion of the embryo.

Parts of Decidua

The decidua can be organized into separate sections, although they have the same composition:

- Decidua basalis: This is the part of the decidua which is located basalolateral to the embryo after implantation.

- Decidua capsularis: Decidua capsularis grows over the embryo on the luminal side, enclosing it into the endometrium. It surrounds the embryo together with decidua basalis.

- Decidua parietalis: All other decidua on the uterine surface belongs to decidua parietalis.

Decidua Throughout Pregnancy

After implantation the decidua remains, at least through the first trimester. However, its most prominent time is during the early stages of pregnancy, during implantation. Its function as a surrounding tissue is replaced by the definitive placenta. However, some elements of the decidualization remain throughout pregnancy.

The compacta and spongiosa layers are still observable beneath the decidua in pregnancy. The glands of the spongiosa layer continue to secrete during the first trimester, when they degenerate. However, before that disappearance, some glands secrete unequally much. This phenomenon of hypersecretion is called the Arias-Stella phenomenon, after the pathologist Javier Arias-Stella.

Pinopodes

Pinopodes are small, finger-like protrusions from the endometrium. They appear between day 19 and day 21 of gestational age. This corresponds to a fertilization age of approximately five to seven days, which corresponds well with the time of implantation. They only persist for two to three days. The development of them is enhanced by progesterone but inhibited by estrogens.

Function in Implantation

Pinopodes endocytose uterine fluid and macromolecules in it. By doing so, the volume of the uterus decreases, taking the walls closer to the embryoblast floating in it. Thus, the period of active pinocytes might also limit the implantation window.

Function during Implantation

Pinopodes continue to absorb fluid, and removes most of it during the early stages of implantation.

Adaptation of Secretions

Not only the lining of the uterus transforms, but in addition, the secretion from its epithelial glands changes. This change is induced by increased levels of progesterone from the corpus luteum. The target of the secretions is the embryoblast, and has several functions on it.

Nourishment

The embryoblast spends approximately 72 hours in the uterine cavity before implanting. In that time, it cannot receive nourishment directly from the blood of the mother, and must rely on secreted nutrients into the uterine cavity, e.g. iron and fat-soluble vitamins.

Growth and Implantation

In addition to nourishment, the endometrium secretes several steroid-dependent proteins, important for growth and implantation. Cholesterol and steroids are also secreted. Implantation is further facilitated by synthesis of matrix substances, adhesion molecules and surface receptors for the matrix substances.

Mechanism

Implantation is initiated when the blastocyst comes into contact with the uterine wall.

Zona Hatching

To be able to perform implantation, the blastocyst first needs to get rid of its zona pellucida. This process can be called "hatching".

Factors

Lytic factors in the uterine cavity, as well as factors from the blastocyst itself are essential for this process. Mechanisms in the latter are indicated by that the zona pellucida remains intact if an unfertilized egg is placed in the uterus under the same conditions. A substance probably involved is plasmin. Plasminogen, the plasmin precursor, is found in the uterine cavity, and blastocyst factors contribute to its conversion to active plasmin. This hypothesis is supported by lytic effects in vitro by plasmin. Furthermore, plasmin inhibitors also inhibit the entire zona hatching in rat experiments.

Apposition

The very first, albeit loose, connection between the blastocyst and the endometrium is called the apposition.

Location

On the endometrium, the apposition is usually made where there is a small crypt in it, perhaps because it increases the area of contact with the rather spherical blastocyst.

On the blastocyst, on the other hand, it occurs at a location where there has been enough lysis of the zona pellucida to have created a rupture to enable direct contact between the underlying trophoblast and the decidua of the endometrium. However, ultimately, the inner cell mass, inside the trophoblast layer, is aligned closest to the decidua. Nevertheless, the apposition on the blastocyst is not dependent on if it is on the same side of the blastocyst as the inner cell mass. Rather, the inner cell mass rotates inside the trophoblast to align to the apposition. In short, the entire surface of the blastocyst has a potential to form the apposition to the decidua.

Molecular Mechanism

The identity of the molecules on the trophoblast and the endometrial epithelia that mediate the initial interaction between the two remain unidentified. However, a number of research groups have proposed that MUC1, a member of the Mucin family of glycosylated proteins, is involved. MUC1 is a transmembrane glycoprotein expressed at the apical surface of endometrial epithelial cells during the window of implantation in humans and has been shown to be differentially expressed between fertile and infertile subjects during this time. MUC1 displays carbohydrate moieties on its extracellular domain that are ligands of L-selectin, a protein expressed on the surface of trophoblast cells. An in vitro model of implantation developed by Genbacev *et al.*, gave evidence to support the hypothesis that L-selectin mediates apposition of the blastocyst to the uterine epithelium by interacting with its ligands.

Adhesion

Adhesion is a much stronger attachment to the endometrium than the loose apposition. The trophoblasts adhere by penetrating the endometrium, with protrusions of trophoblast cells.

Communication

There is massive communication between the blastocyst and the endometrium at this stage. The blastocyst signals to the endometrium to adapt further to its presence, e.g. by changes in the cytoskeleton of decidual cells. This, in turn, dislodges the decidual cells from their connection to the underlying basal lamina, which enables the blastocyst to perform the succeeding invasion.

This communication is conveyed by receptor-ligand-interactions, both integrin-matrix and proteoglycan ones.

Proteoglycan Receptors

Another ligand-receptor system involved in adhesion is proteoglycan receptors, found on the surface of the decidua of the uterus. Their counterparts, the proteoglycans, are found around the trophoblast cells of the blastocyst. This ligand-receptor system also is present just at the implantation window.

Invasion

Invasion is an even further establishment of the blastocyst in the endometrium.

Syncytiotrophoblasts

The protrusions of trophoblast cells that adhere into the endometrium continue to proliferate and penetrate into the endometrium. As these trophoblast cells penetrate, they differentiate to become a new type of cells, syncytiotrophoblast. The prefix syn- refers to the transformation that occurs as the boundaries between these cells disappear to form a single mass of many cell nuclei (a syncytium). The rest of the trophoblasts, surrounding the inner cell mass, are hereafter called cytotrophoblasts.

Invasion continues with the syncytiotrophoblasts reaching the basal membrane beneath the decidual cells, penetrating it and further invading into the uterine stroma. Finally, the whole embryo is embedded in the endometrium. Eventually, the syncytiotrophoblasts come into contact with maternal blood and form chorionic villi. This is the initiation of forming the placenta.

Secretions

The blastocyst secretes factors for a multitude of purposes during invasion. It secretes several autocrine factors, targeting itself and stimulating it to further invade the endometrium. Furthermore, secretions loosen decidual cells from each other, prevent the embryo from being rejected by the mother, trigger the final decidualization and prevent menstruation.

Autocrine

Human chorionic gonadotropin is an autocrine growth factor for the blastocyst. Insulin-like growth factor 2, on the other hand, stimulates the invasiveness of it.

Dislodging

The syncytiotrophoblasts dislodges decidual cells in their way, both by degradation of cell adhesion molecules linking the decidual cells together as well as degradation of the extracellular matrix between them.

Cell adhesion molecules are degraded by syncytiotrophoblast secretion of Tumor necrosis factor-alpha. This inhibits the expression of cadherins and beta-catenin. Cadherins are cell adhesion molecules, and beta-catenin helps to anchor them to the cell membrane. Inhibited expression of these molecules thus loosens the connection between decidual cells, permitting the syncytotrophoblasts and the whole embryo with them to invade into the endometrium.

The extracellular matrix is degraded by serine endopeptidases and metalloproteinases. Examples of such metalloproteinases are collagenases, gelatinases and stromelysins. These collagenases digest Type-I collagen, Type-II collagen, Type-III collagen, Type-VII collagen and Type-X collagen. The gelatinases exist in two forms; one digesting Type-IV collagen and one digesting gelatin.

Immunosuppressive

The embryo differs from the cells of the mother, and would be rejected as a parasite by the immune system of the mother if it didn't secrete immunosuppressive agents. Such agents are Platelet-activating factor, human chorionic gonadotropin, early pregnancy factor, immunosuppressive factor, Prostaglandin E2, Interleukin 1-alpha, Interleukin 6, interferon-alpha, leukemia inhibitory factor and Colony-Stimulating Factor.

Decidualization

Factors from the blastocyst also trigger the final formation of decidual cells into their proper form. In contrast, some decidual cells in the proximity of the blastocyst degenerate, providing nutrients for it.

Prevention of Menstruation

Human chorionic gonadotropin (hCG) not only acts as an immunosuppressive, but also "notifies" the mother's body that she is pregnant, preventing menstruation by sustaining the function of the corpus luteum.

Failure

Implantation failure is considered to be caused by inadequate uterine receptivity in two-thirds of cases, and by problems with the embryo itself in the other third.

Inadequate uterine receptivity may be caused by abnormal cytokine and hormonal signaling as well as epigenetic alterations. Recurrent implantation failure is a cause of female infertility. Therefore, pregnancy rates can be improved by optimizing endometrial receptivity for implantation. Evaluation of implantation markers may help to predict pregnancy outcome and detect occult implantation deficiency.

Luteal support is the administration of medication, generally progestins, for the purpose of increasing the success rate of implantation and early embryogenesis, thereby complementing the function of the corpus luteum.

Embryonic Disc

The floor of the amniotic cavity is formed by the embryonic disc (or embryonic disk) composed of a layer of prismatic cells, the embryonic ectoderm, derived from the inner cell-mass and lying in apposition with the endoderm.

In humans, it is the stage of development that occurs after implantation and prior to the embryonic folding (e.g. seen between about day 14 to day 21 post fertilization). It is derived from the epiblast layer, which lies between the hypoblast layer and the amnion. The epiblast layer is derived from the inner cell mass. Through the process of gastrulation, the bilaminar embryonic disc becomes trilaminar. The notochord forms thereafter. Through the process of neurulation, the notochord induces the formation of the neural tube in the embryonic disc.

Gastrulation

Gastrulation is a phase early in the embryonic development of most animals, during which the single-layered blastula is reorganized into a multilayered structure known as the gastrula. Before gastrulation, the embryo is a continuous epithelial sheet of cells; by the end of gastrulation, the embryo has begun differentiation to establish distinct cell lineages, set up the basic axes of the body (e.g. dorsal-ventral, anterior-posterior), and internalized one or more cell types including the prospective gut.

In triploblastic organisms the gastrula is trilaminar ("three-layered"). These three *germ layers* are known as the ectoderm, mesoderm, and endoderm. In diploblastic organisms, such as Cnidaria and Ctenophora, the gastrula has only ectoderm and endoderm. The two layers are also sometimes referred to as the hypoblast and epiblast.

Gastrulation takes place after cleavage and the formation of the blastula. Gastrulation is followed by organogenesis, when individual organs develop within the newly formed germ layers. Each layer gives rise to specific tissues and organs in the developing embryo. The ectoderm gives rise to epidermis, the nervous system, and to the neural crest in vertebrates. The endoderm gives rise to the epithelium of the digestive system and respiratory system, and organs associated with the digestive system, such as the liver and pancreas. The mesoderm gives rise to many cell types such as muscle, bone, and connective tissue. In vertebrates, mesoderm derivatives include the noto-chord, the heart, blood and blood vessels, the cartilage of the ribs and vertebrae, and the dermis. Following gastrulation, cells in the body are either organized into sheets of connected cells (as in epithelia), or as a mesh of isolated cells, such as mesenchyme.

The molecular mechanism and timing of gastrulation is different in different organisms. However, some common features of gastrulation across triploblastic organisms include: (1) A change in the topological structure of the embryo, from a simply connected surface (sphere-like), to a non-simply connected surface (torus-like); (2) the differentiation of cells into one of three types (endodermal, mesodermal, and ectodermal); and (3) the digestive function of many endodermal cells. The signaling pathways, which refers to the signals that indicate activation or inhibition of something else in the organism, are often different depending on the organism as well.

Lewis Wolpert, pioneering developmental biologist in the field, has been credited for noting that "It is not birth, marriage, or death, but gastrulation which is truly the most important time in your life."

The terms "gastrula" and "gastrulation" were coined by Ernst Haeckel, in his 1872 work "Biology of Calcareous Sponges".

Although gastrulation patterns exhibit enormous variation throughout the animal kingdom, they are unified by the five basic types of cell movements that occur during gastrulation: 1) invagination 2) involution 3) ingression 4) delamination 5) epiboly.

Classic Model Systems

Gastrulation is highly variable across the animal kingdom but has underlying similarities. Gastrulation has been studied in many animals, but some models have been used for longer than others. Furthermore, it is easier to study development in animals that develop outside the mother. Animals whose gastrulation is understood in the greatest detail include:

- Mollusc
- Sea urchin

- Frog

- Chicken

Protostomes versus Deuterostomes

The distinction between protostomes and deuterostomes is based on the direction in which the mouth (stoma) develops in relation to the blastopore. Protostome derives from the word protostoma meaning "first mouth"whereas Deuterostome's etymology is "second mouth" from the words second and mouth.

The major distinctions between deuterostomes and protostomes are found in embryonic development:

- Mouth/anus:

  ○ In protostome development, the first opening in development, the blastopore, becomes the animal's mouth.

  ○ In deuterostome development, the blastopore becomes the animal's anus.

- Cleavage:

  ○ Protostomes have what is known as spiral cleavage which is determinate, meaning that the fate of the cells is determined as they are formed.

  ○ Deuterostomes have what is known as radial cleavage that is indeterminate.

Sea Urchins

Sea urchins Euechinoidea have been an important model system in developmental biology since the 19th century. Their gastrulation is often considered the archetype for invertebrate deuterostomes.

Germ Layer Determination

Sea urchins exhibit highly stereotyped cleavage patterns and cell fates. Maternally deposited mRNAs establish the organizing center of the sea urchin embryo. Canonical Wnt and Delta-Notch signaling progressively segregate progressive endoderm and mesoderm.

Cell Internalization

In Euechinoids the first cells to internalize are the primary mesenchyme cells (PMCs), which have a skeletogenic fate, which ingress during the blastula stage. Gastrulation – internalization of the prospective endoderm and non-skeletogenic mesoderm – begins shortly thereafter with invagination and other cell rearrangements the vegetal pole, which contribute approximately 30% to the final archenteron length. The gut's final length depends on cell rearrangements within the archenteron.

Amphibians

Tailless amphibians (Anura) are a classic model system for gastrulation.

Symmetry Breaking

The sperm contributes one of the two mitotic asters needed to complete first cleavage. The sperm can enter anywhere in the animal half of the egg but its exact point of entry will break the egg's radial symmetry by organizing the cytoskeleton. Prior to first cleavage, the egg's cortex rotates relative to the internal cytoplasm by the coordinated action of microtubules, in a process known as cortical rotation. This displacement brings maternally loaded determinants of cell fate from the equatorial cytoplasm and vegetal cortex into contact, and together these determinants set up the organizer. Thus, the area on the vegetal side opposite the sperm entry point will become the organizer. Hilde Mangold, working in the lab of Hans Spemann, demonstrated that this special "organizer" of the embryo is necessary and sufficient to induce gastrulation.

Germ Layer Determination

Specification of endoderm depends on rearrangement of maternally deposited determinants, leading to nuclearization of Beta-catenin. Mesoderm is induced by signaling from the presumptive endoderm to cells that would otherwise become ectoderm.

Cell Internalization

The dorsal lip of the blastopore is the mechanical driver of gastrulation. The first sign of invagination seen in this video of frog gastrulation is the dorsal lip.

Cell Signaling

In the frog, *Xenopus,* one of the signals is retinoic acid (RA). RA signaling in this organism can affect the formation of the endoderm and depending on the timing of the signaling, it can determine the fate whether its pancreatic, intestinal, or respiratory. Other signals such as Wnt and BMP also play a role in respiratory fate of the *Xenopus* by activating cell lineage tracers.

Amniotes

In amniotes (reptiles, birds and mammals), gastrulation involves the creation of the blastopore, an opening into the archenteron. Note that the blastopore is not an opening into the blastocoel, the space within the blastula, but represents a new inpocketing that pushes the existing surfaces of the blastula together. In amniotes, gastrulation occurs in the following sequence: (1) the embryo becomes asymmetric; (2) the primitive streak forms; (3) cells from the epiblast at the primitive streak undergo an epithelial to mesenchymal transition and ingress at the primitive streak to form the germ layers.

Symmetry Breaking

In preparation for gastrulation, the embryo must become asymmetric along both the proximal-distal axis and the anterior-posterior axis. The proximal-distal axis is formed when the cells of the embryo form the "egg cylinder," which consists of the extraembryonic tissues, which give rise to structures like the placenta, at the proximal end and the epiblast at the distal end. Many signaling pathways contribute to this reorganization, including BMP, FGF, nodal, and Wnt. Visceral endoderm surrounds the epiblast. The distal visceral endoderm (DVE) migrates to the anterior portion of the embryo, forming the "anterior visceral endoderm" (AVE). This breaks anterior-posterior symmetry and is regulated by nodal signaling.

Germ Layer Determination

The primitive streak is formed at the beginning of gastrulation and is found at the junction between the extraembryonic tissue and the epiblast on the posterior side of the embryo and the site of ingression. Formation of the primitive streak is reliant upon nodal signaling in the Koller's sickle within the cells contributing to the primitive streak and BMP4 signaling from the extraembryonic tissue. Furthermore, Cer1 and Lefty1 restrict the primitive streak to the appropriate location by antagonizing nodal signaling. The region defined as the primitive streak continues to grow towards the distal tip.

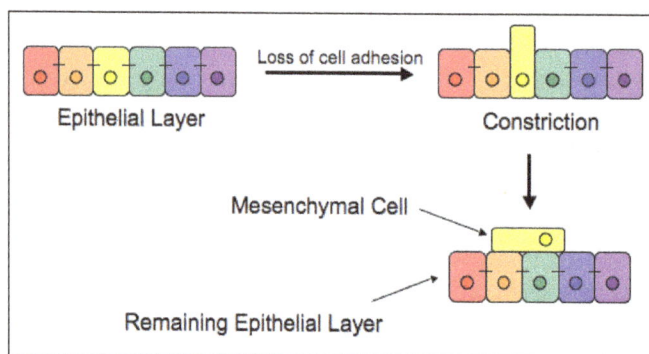

Epithelial to Mesenchmyal Cell Transition – loss of cell adhesion leads to constriction and extrusion of newly mesenchymal cell.

During the early stages of development, the primitive streak is the structure that will establish bilateral symmetry, determine the site of gastrulation and initiate germ layer formation. To form the streak, reptiles, birds and mammals arrange mesenchymal cells along the prospective midline, establishing the first embryonic axis, as well as the place where cells will ingress and migrate during the process of gastrulation and germ layer formation. The primitive streak extends through this midline and creates the antero-posterior body axis, becoming the first symmetry-breaking event in the embryo, and marks the beginning of gastrulation. This process involves the ingression of mesoderm and endoderm progenitors and their migration to their ultimate position, where they will differentiate into the three germ layers. The localization of the cell adhesion and signaling molecule beta-catenin is critical to the proper formation of the organizer region that is responsible for initiating gastrulation.

Cell Internalization

In order for the cells to move from the epithelium of the epiblast through the primitive streak to form a new layer, the cells must undergo an epithelial to mesenchymal transition (EMT) to lose their epithelial characteristics, such as cell-cell adhesion. FGF signaling is necessary for proper EMT. FGFR1 is needed for the up regulation of SNAI1, which down regulates E-cadherin, causing a loss of cell adhesion. Following the EMT, the cells ingress through the primitive streak and spread out to form a new layer of cells or join existing layers. FGF8 is implicated in the process of this dispersal from the primitive streak.

Cell Signaling

There are certain signals that play a role in determination and formation of the three germ layers, such as FGF, RA, and Wnt. In mammals such as mice, RA signaling can play a role in lung formation. If there isn't enough RA, there will be an error in the lung production. RA also regulates the respiratory competence in this mouse model.

Cell Signaling Driving Gastrulation

During gastrulation, the cells are differentiated into the ectoderm or mesendoderm, which then separates into the mesoderm and endoderm. The endoderm and mesoderm form due to the nodal signaling. Nodal signaling uses ligands that are part of TGFβ family. These ligands will signal transmembrane serine/threonine kinase receptors, and this will then phosphorylate Smad2 and Smad3. This protein will then attach itself to Smad4 and relocate to the nucleus where the mesendoderm genes will begin to be transcribed. The Wnt pathway along with β-catenin plays a key role in nodal signaling and endoderm formation. Fibroblast growth factors (FGF), canonical Wnt pathway, bone morphogenetic protein (BMP), and retinoic acid (RA) are all important in the formation and development of the endoderm. FGF are important in producing the homeobox gene which regulates early anatomical development. BMP signaling plays a role in the liver and promotes hepatic fate. RA signaling also induce homeobox genes such as Hoxb1 and Hoxa5. In mice, if there is a lack in RA signaling the mouse won't develop lungs. RA signaling also has multiple uses in organ formation of the pharyngeal arches, the foregut, and hindgut.

Gastrulation in Vitro

There have been a number of attempts to understand the processes of Gastrulation using *in vitro* techniques in parallel and complementary to studies in embryos, usually though the use of 2D and 3D cell (Embryonic organoids) culture techniques using Embryonic stem cells (ESCs) or induced pluripotent stem cells (iPSCs). These are associated with number of clear advantages in using tissue-culture based protocols, some of which include reducing the cost of associated *in vivo* work (thereby reducing, replacing and refining the use of animals in experiments; the 3Rs), being able to accurately apply agonists/antagonists in spatially and temporally specific manner which may be technically difficult to perform during Gastrulation. However, it is important to relate the observations in culture to the processes occurring in the embryo for context.

To illustrate this, the guided differentiation of mouse ESCs has resulted in generating primitive streak-like cells that display many of the characteristics of epiblast cells that traverse through the primitive streak (e.g. transient brachyury up regulation and the cellular changes associated with an epithelial to mesenchymal transition), and human ESCs cultured on micro patterns, treated with BMP4, can generate spatial differentiation pattern similar to the arrangement of the germ layers in the human embryo. Finally, using 3D embryoid body- and organoid-based techniques, small aggregates of mouse ESCs (Embryonic Organoids, or Gastruloids) are able to show a number of processes of early mammalian embryo development such as symmetry-breaking, polarisation of gene expression, gastrulation-like movements, axial elongation and the generation of all three embryonic axes (anteroposterior, dorsoventral and left-right axes).

Neurulation

Neurulation and Formation of the Spinal Cord

Neurulation is the process of formation of the hollow neural tube by folding of the epithelial neural plate . Neurulation in humans occurs in two distinct phases: primary neurulation during weeks 3 and 4 of gestation leading to development of the brain and spinal cord , and secondary neurulation during weeks 5 and 6, with formation of the lower sacral and coccygeal cord. Defects of neurulation are the earliest abnormalities of brain development that are clinically detectable in fetal life and extend into postnatal life.

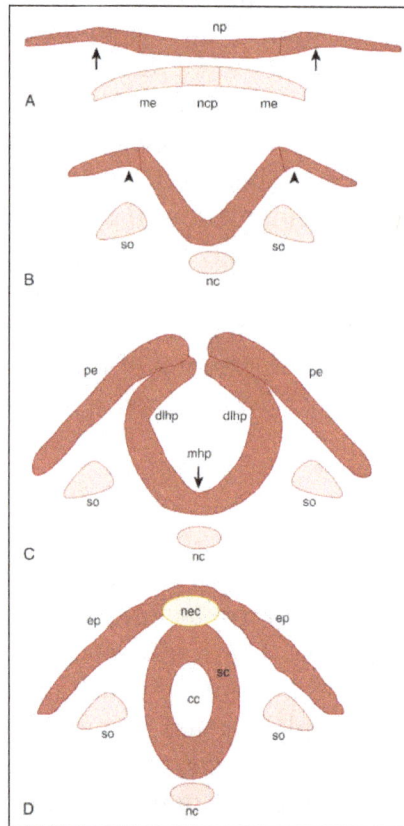

A simplified diagram of neural tube closure.

Neural folds form at the lateral extremes of the neural plate (A, arrows), elevate (B, arrows) and converge toward the dorsal midline (C), and fuse at their dorsal tips to form the closed neural tube (D). Bending or hinge points form at two sites: the median hinge point (mhp) overlying the notochord and the paired dorsolateral hinge points (dlhp) at the lateral sides of the folds cc, Central canal; ep, epidermis; me, mesoderm; nc, notochord; ncp, notochordal plate; nec, neural crest; np, neural plate; pe, presumptive epidermis; sc, spinal cord; so, somites.

The neural plate elongates into a drop-shaped structure, broader at its cranial end and narrower in the future spinal regions. This morphogenetic event, known as convergent extension, is under the control of the noncanonical Wnt pathway and downstream proteins such as VANGL1, CELSR1, SCRB1, and Dishevelled.

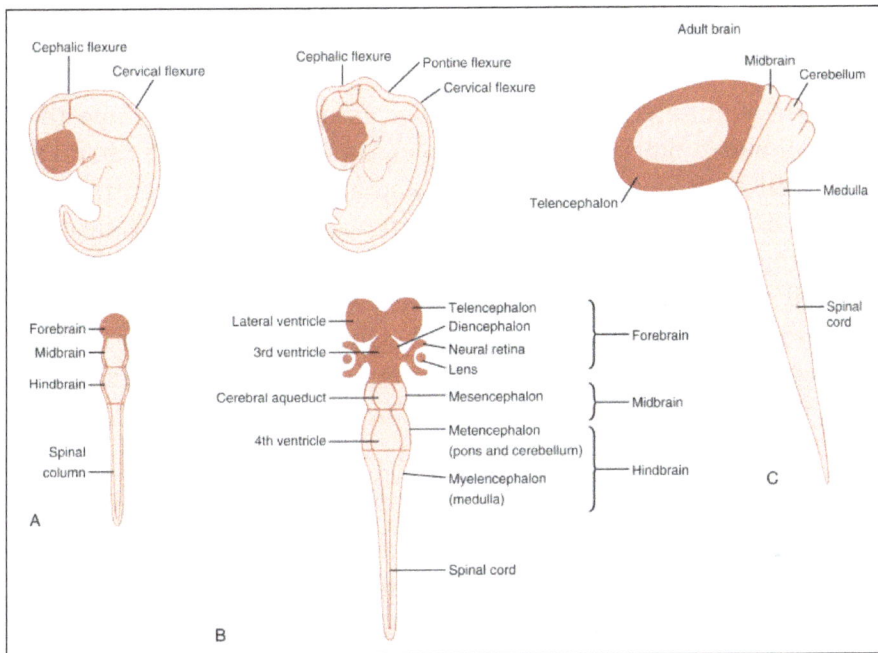

Adult brain

Cephalic flexure
Cervical flexure
Cephalic flexure Pontine flexure
Cervical flexure
Midbrain
Cerebellum
Telencephalon
Medulla

Forebrain
Midbrain
Hindbrain
Lateral ventricle
3rd ventricle
Cerebral aqueduct
4th ventricle
Telencephalon
Diencephalon
Neural retina
Lens
Mesencephalon
Metencephalon
(pons and cerebellum)
Myelencephalon
(medulla)
Forebrain
Midbrain
Hindbrain
Spinal
cord
C

Spinal
column
A
Spinal cord
B

Development of the brain and spinal cord from the neural tube.

The neural plate is further shaped by bending at the median hinge point overlying the notochord at the future upper spinal cord and at the paired dorsolateral hinge points at the levels of the brain and lower spinal cord. This differential bending appears to be under the control of signals diffusing from the notochord, including the signal transduction molecule Sonic hedgehog (SHH), which is also the major determinant of ventral neural progenitor identity in embryonic spinal cord and forebrain. Differential gradients of expression between the ventral SHH expression and BMPs in the dorsal ectoderm not only establish a dorsoventral plane but also lead to the later establishment of distinct classes of neurons with the spinal cord. The neural plate rotates by elevation and convergence around the median and dorsal hinge points. This bending appears to be dependent on apical constriction of columnar neural tube cells to become wedge shaped under the control of actin-related genes, such as *ARHGAP35* (a negative regulator of Rho GTPase), *MARCKS* (a protein kinase C target), shroom genes (which encode an actin-associated protein family), and *VCL* (which encodes the actin-binding protein vinculin). Cytoskeletal proteins appear to regulate only cranial neurulation, whereas the spinal neural tube closes normally despite defects in several cytoskeleton-related genes.

Neural tube closure appears to continue by the development of bilateral folds at its junction with nonneural ectoderm. These folds elevate and become opposed in the midline, with fusion occurring by interdigitation of cellular protrusions from apical cells and the formation of permanent cell contacts. Membrane-bound ephrin ligands and their Eph tyrosine kinase receptors have been implicated in this process. Maintenance of proliferation by Notch signaling and apoptotic programmed cell death play important but poorly understood roles in the process as well.

Neural tube closure is a discontinuous process occurring at two invariant sites in humans, with a third variable site representing a potential factor for neural tube defects (NTDs). The open neural folds between initiation closure sites are known as neuropores. As closure progresses, the neuropores gradually shorten and close, leading to an intact closed neural tube. This is subsequently covered by ectoderm-derived epidermis.

Secondary neurulation is initiated after primary neurulation is complete and the posterior neuropore closes. The tail bud, a pluripotent mass of cells, a remnant of the caudal primitive streak, proliferates and condenses, followed by cavitation and fusion with the central canal of the neural tube. As this process of canalization progresses during ensuing weeks, neurons and ependymal cells differentiate to form the caudal end of the spinal cord. Resorption of the tail bud and other cells of the caudal cell mass leaves the filum terminale, which often contains ependymal cell nests along its length. These nests of ependymal cells can begin to proliferate later in life as a monoclonal population of glial cells. These ependymomas, uncommon glial cell tumors, are among the most common tumors of the cauda equina and filum terminale.

The closed neural tube consists of a thick pseudostratified epithelium of neuroepithelial cells that begin to divide rapidly immediately after closure and give rise to a second cell type, the neuroblast. These cells form the mantle layer, the future gray matter of the spinal cord. This further gives rise to an outermost layer, the marginal layer that will become myelinated and form the future white matter of the spinal cord. As neuroblast proliferation continues, the neural tube develops dorsal and ventral thickenings that become the alar and basal plates that will give rise to the sensory and motor areas of the spinal cord, respectively. A longitudinal groove, the sulcus limitans, demarcates the boundary between the two.

As the neural tube closes, cells at the edge of the neural plate separate from the neural epithelium and migrate into the extracellular matrix to become neural crest cells. Neural crest migration is required for complete closure of the cranial neural tube but not for spinal neural tube closure. Although cell adhesion molecules were previously thought to play a role in neurulation, more recent studies have shown mice with null mutations in neural cell adhesion molecule or N-cadherin undergo normal neurulation. Down-regulation of cell adhesion molecules, transition from tight junction to gap junction molecules, and an increased expression of matrix metalloproteinases play an important role in neural crest migration.

Neural crest cells migrate widely to become neurons and glia in dorsal root and autonomic ganglia. The fate of neural crest–derived cells is also to provide innervation for the gastrointestinal tract and to become neurons in the sensory ganglia of the cranial nerves (also formed in part by ectodermal placodes) and to become melanocytes, cartilage of the bone and face, and a variety of endocrine and structural tissues.

The ultimate phenotypic and developmental fate of neural crest cells is tied to the timing, mode, and pattern of migration and is in large part controlled by the environment in which they reside. In addition to various permissive versus inhibitory signaling molecules including Eph/ephrin, semaphorin/neuropilin, and Robo/Slit, a complex relationship exists between neural crest cells and the developing vasculature. The dorsal aorta, the first major blood vessel established during embryogenesis, expresses BMP signals to direct migration and lineage segregation of neural crest cells into adrenal medulla or sympathetic ganglia. Schwann cells, neural crest–derived glia that ensheath peripheral nervous system neurons, direct patterning of arterial vasculature parallel to sensory nerves through the expression of *CXCL12* (also known as *SDF1*). Neural crest cells retain a relatively broad developmental potential as they begin migration and their ultimate fate is strongly influenced by local factors.

References

- Bernstein H and Bernstein C (2013). Evolutionary Origin and Adaptive Function of Meiosis. In Meiosis: Bernstein C and Bernstein H, editors, intech. ISBN 978-953-51-1197-9

- What are yeasts?". Yeast Virtual Library. 13 September 2009. Archived from the original on 26 February 2009. Retrieved 28 November 2009

- Bernstein, Harris; Bernstein, Carol; Michod, Richard E. (2011). "Meiosis as an Evolutionary Adaptation for DNA Repair". In Kruman, Inna (ed.). DNA Repair. Doi:10.5772/25117. ISBN 978-953-307-697-3

- Gilbert, Scott F. (2000). "Early Development of the Nematode Caenorhabditis elegans". Developmental Biology (6th ed.). Retrieved 2007-09-17

- Animal-development, science: britannica.com, Retrieved 14 July, 2019

- Gilbert, Scott (2010). Developmental Biology 9th Ed + Devbio Labortatory Vade Mecum3. Sinauer Associates Inc. Pp. 243–247, 161. ISBN 978-0-87893-558-1

- Neurulation, immunology-and-microbiology: sciencedirect.com, Retrieved 17 May, 2019

Development Processes in Plants

4

- **Plant Reproduction**

- **Development and Growth**

- **Photomorphogenesis**

- **Morphological Variation**

- **Plant Evolutionary Developmental Biology**

Plant development refers to the production of new tissues and structures by plants throughout their life from meristems. Some of the most important processes of plant development are plant reproduction, photomorphogenesis, morphological variation, etc. This chapter discusses in detail these processes related to plant development.

Plant Reproduction

Reproduction in plants is either asexual or sexual. Asexual reproduction in plants involves a variety of widely disparate methods for producing new plants identical in every respect to the parent. Sexual reproduction, on the other hand, depends on a complex series of basic cellular events, involving chromosomes and their genes, that take place within an elaborate sexual apparatus evolved precisely for the development of new plants in some respects different from the two parents that played a role in their production.

In order to describe the modification of reproductive systems, plant groups must be identified. One convenient classification of organisms sets plants apart from other forms such as bacteria, algae, fungi, and protozoans. Under such an arrangement, the plants, as separated, comprise two great divisions (or phyla)—the Bryophyta (mosses and liverworts) and the Tracheophyta (vascular plants). The vascular plants include four subdivisions: the three entirely seedless groups are the Psilopsida, Lycopsida, and Sphenopsida; the fourth group, the Pteropsida, consists of the ferns (seedless) and the seed plants (gymnosperms and angiosperms).

A comparative treatment of the two patterns of reproductive systems will introduce the terms required for an understanding of the survey of those systems as they appear in selected plant groups.

General Features of Asexual Systems

Asexual reproduction involves no union of cells or nuclei of cells and, therefore, no mingling of genetic traits, since the nucleus contains the genetic material (chromosomes) of the cell. Only those systems of asexual reproduction that are not really modifications of sexual reproduction are considered below. They fall into two basic types: systems that utilize almost any fragment or part of a plant body and systems that depend upon specialized structures that have evolved as reproductive agents.

Reproduction by Fragments

In many plant groups, fragmentation of the plant body, followed by regeneration and development of the fragments into whole new organisms, serves as a reproductive system. Fragments of the plant bodies of liverworts and mosses regenerate to form new plants. In nature and in laboratory and greenhouse cultures, liverworts fragment as a result of growth; the growing fragments separate by decay at the region of attachment to the parent. During prolonged drought, the mature portions of liverworts often die, but their tips resume growth and produce a series of new plants from the original parent plant.

Archegonium: The female reproductive organ, or archegonium, emerging from a thalloid liverwort.

In mosses, small fragments of the stems and leaves (even single cells of the latter) can, with sufficient moisture and under proper conditions, regenerate and ultimately develop into new plants.

It is common horticultural practice to propagate desirable varieties of garden plants by means of plant fragments, or cuttings. These may be severed leaves or portions of roots or stems, which are stimulated to develop roots and produce leafy shoots. Naturally fallen branches of willows (Salix) and poplars (Populus) root under suitable conditions in nature and eventually develop into trees. Other horticultural practices that exemplify asexual reproduction include budding (the removal of buds of one plant and their implantation on another) and grafting (the implantation of small branches of one individual on another).

Reproduction by Special Asexual Structures

Throughout the plant kingdom, specially differentiated or modified cells, groups of cells, or organs have, during the course of evolution, come to function as organs of asexual reproduction. These structures are asexual in that the individual reproductive agent develops into a new individual without the union of sex cells (gametes).

Airborne spores characterize most nonflowering land plants, such as mosses, liverworts, and ferns. Although the spores arise as products of meiosis, a cellular event in which the number of chromosomes in the nucleus is halved, such spores are asexual in the sense that they may grow directly into new individuals, without prior sexual union.

Among liverworts, mosses, lycopods, ferns, and seed plants, few-to many-celled specially organized buds, or gemmae, also serve as agents of asexual reproduction.

The vegetative, or somatic, organs of plants may, in their entirety, be modified to serve as organs of reproduction. In this category belong such flowering-plant structures as stolons, rhizomes, tubers, corms, and bulbs, as well as the tubers of liverworts, ferns, and horsetails, the dormant buds of certain moss stages, and the leaves of many succulents. Stolons are elongated runners, or horizontal stems, such as those of the strawberry, which root and form new plantlets when they make proper contact with a moist soil surface. Rhizomes, as seen in iris, are fleshy, elongated, horizontal stems that grow within or upon the soil. The branching of rhizomes results in multiplication of the plant. The enlarged fleshy tips of subterranean rhizomes or stolons are known as tubers, examples of which are potatoes. Tubers are fleshy storage stems, the buds ("eyes") of which, under proper conditions, can develop into new individuals. Erect, vertical, fleshy, subterranean stems, which are known as corms, are exemplified by crocuses and gladioli. These organs tide the plants over periods of dormancy and may develop secondary cormlets, which give rise to new plantlets. Unlike the corm, only a small portion of the bulb, as in lilies and the onion, represents stem tissue. The latter is surrounded by the fleshy food-storage bases of earlier-formed leaves. After a period of dormancy, bulbs develop into new individuals. Large bulbs produce secondary bulbs through development of buds, resulting in an increase in number of individuals.

General Features of Sexual Systems

In most plant groups, both sexual and asexual methods of reproduction occur. Some species, however, seem secondarily to have lost the capacity for sexual reproduction.

The Cellular Basis

Sexual reproduction at the cellular level generally involves the following phenomena: the union of sex cells and their nuclei, with concomitant association of their chromosomes, which contain the genes, and the nuclear division called meiosis. The sex cells are called gametes, and the product of their union is a zygote. All gametes are normally haploid (having a single set of chromosomes) and all zygotes, diploid (having a double set of chromosomes, one set from each parent). Gametes may be motile, by means of whiplike hairs (flagella) or of flowing cytoplasm (amoeboid motion). In their union, gametes may be morphologically indistinguishable (i.e., isogamous) or they may be distinguishable only on the criterion of size (i.e., heterogamous). The larger gamete, or egg, is nonmotile; the smaller gamete, or sperm, is motile. The last type of gametic difference, egg and sperm, is often designated as oogamy. In oogamous reproduction, the union of sperm and egg is called fertilization. Isogamy, heterogamy, and oogamy are often considered to represent an increasingly specialized evolutionary series.

In bryophytes (mosses and liverworts) and tracheophytes (vascular plants)—sexual reproduction is of the oogamous type, or a modification thereof, in which the sex cells, or gametes, are of two types, a larger nonmotile egg and a smaller motile sperm. These gametes are often produced in special containers called gametangia, which are multicellular. In cases in which special gametangia are lacking, every cell produces a gamete. In oogamy, the male gametangia are called antheridia and the female oogonia or archegonia. A female gametangium with a sterile cellular jacket is called an archegonium, although, like an oogonium, it produces eggs.

The Plant Basis

Individual plants may be either bisexual (hermaphroditic), in which male and female gametes are produced by the same organism, or unisexual, producing either male or female gametes but not both. A bisexual individual, however, is not necessarily capable of fertilizing its own eggs. In certain ferns, for example, male gametes of one individual are not compatible with the female gametes of the same individual, so cross-fertilization (with another individual of the species) is obligatory. This situation, of course, is similar in adaptive significance to cross-pollination (which leads to cross-fertilization) among seed plants.

Among the liverworts, mosses, and vascular plants, the life cycle involves two different phases, often called generations, although only one plant generation is, in fact, involved in one complete cycle. This type of life cycle is often said to illustrate the "alternation of generations," in which a haploid individual (i.e., with one set of chromosomes), or tissue, called a gametophyte, at maturity produces gametes that unite in pairs to form diploid (i.e., containing two sets of chromosomes) zygotes. The latter develop directly into individuals, or tissues, called sporophytes, in which the nuclei of certain fertile cells, called spore mother cells, or sporocytes, give rise to haploid spores (sometimes called meiospores). These spores are lightweight and are borne by air currents; they germinate to form the haploid, sexual, gamete-producing phase, usually designated the gametophyte.

There are several variations in the life cycle. The haploid gametophyte and sporophyte may be free-living, independent plants (e.g., certain algae and yeasts), in which case the life cycle is diplobiontic, or the sporophyte may be physically attached to the gametophyte, as it is in liverworts and mosses. By contrast, the gametophytic phases develop as parasites on the sporophytes of the seed plants, as in certain algae. In further variation, the alternating phases may be similar morphologically except for the type of reproductive cells (gametes or spores) they produce (isomorphic life cycle), or they may be strikingly dissimilar, as in some algae, mosses, ferns, and seed plants (heteromorphic life cycle). Only heteromorphic life cycles occur in liverworts, mosses, vascular plants, and certain fungi.

The differences between the gametophyte and sporophyte are often great, especially those of the diplobiontic types, so the alternates seem to be two different, unrelated individuals rather than different manifestations of the same organism.

Development and Growth

"Development" and "growth" are sometimes used interchangeably in conversation, but in a botanical sense they describe separate events in the organization of the mature plant body.

Development is the progression from earlier to later stages in maturation, e.g. a fertilized egg *develops* into a mature tree. It is the process whereby tissues, organs, and whole plants are produced. It involves: growth, morphogenesis (the acquisition of form and structure), and differentiation. The interactions of the environment and the genetic instructions inherited by the cells determine how the plant develops.

Growth is the irreversible change in size of cells and plant organs due to both cell division *and* enlargement. Enlargement necessitates a change in the elasticity of the cell walls together with an increase in the size and water content of the vacuole. Growth can be determinate—when an organ or part or whole organism reaches a certain size and then stops growing—or indeterminate—when cells continue to divide indefinitely. Plants in general have indeterminate growth.

Differentiation is the process in which generalized cells specialize into the morphologically and physiologically different cells. Since all of the cells produced by division in the meristems have the same genetic make up, differentiation is a function of which particular genes are either expressed orrepressed. The kind of cell that ultimately develops also is a result of its location: Root cells don't form in developing flowers, for example, nor do petals form on roots.

Mature plant cells can be stimulated under certain conditions to divide and differentiate again, i.e. to dedifferentiate. This happens when tissues are wounded, as when branches break or leaves are damaged by insects. The plant repairs itself bydedifferentiating parenchyma cells in the vicinity of the wound, making cells like those injured or else physiologically similar cells.

Plants differ from animals in their manner of growth. As young animals mature, all parts of their bodies grow until they reach a genetically determined size for each species. Plant growth, on the other hand, continues throughout the life span of the plant and is restricted to certain meristematic tissue regions only. This continuous growth results in:

- Two general groups of tissues, primary and secondary.

- Two body types, primary and secondary.

- Apical and lateral meristems.

Apical meristems, or zones of cell division, occur in the tips of both roots and stems of all plants and are responsible for increases in the length of the primary plant body as the primary tissues differentiate from the meristems. As the vacuoles of the primary tissue cells enlarge, the stems and roots increase in girth until a maximum size (determined by the elasticity of their cell walls) is reached. The plant may continue to grow in length, but no longer does it grow in girth. Herbaceous plants with only primary tissues are thus limited to a relatively small size.

Woody plants, on the other hand, can grow to enormous size because of the strengthening and protective secondary tissues produced by lateral meristems, which develop around the periphery of their roots and stems. These tissues constitute the secondary plant body.

Plants produce new tissues and structures throughout their life from meristems located at the tips of organs, or between mature tissues. Thus, a living plant always has embryonic tissues. By contrast, an animal embryo will very early produce all of the body parts that it will ever have in its life. When the animal is born (or hatches from its egg), it has all its body parts and from that point will only grow larger and more mature.

The properties of organization seen in a plant are emergent properties which are more than the sum of the individual parts. "The assembly of these tissues and functions into an integrated multicellular organism yields not only the characteristics of the separate parts and processes but also quite a new set of characteristics which would not have been predictable on the basis of examination of the separate parts."

A vascular plant begins from a single celled zygote, formed by fertilisation of an egg cell by a sperm cell. From that point, it begins to divide to form a plant embryo through the process of embryogenesis. As this happens, the resulting cells will organize so that one end becomes the first root while the other end forms the tip of the shoot. In seed plants, the embryo will develop one or more "seed leaves" (cotyledons). By the end of embryogenesis, the young plant will have all the parts necessary to begin in its life.

Once the embryo germinates from its seed or parent plant, it begins to produce additional organs (leaves, stems, and roots) through the process of organogenesis. New roots grow from root meristems located at the tip of the root, and new stems and leaves grow from shoot meristems located at the tip of the shoot. Branching occurs when small clumps of cells left behind by the meristem, and which have not yet undergone cellular differentiation to form a specialized tissue, begin to grow as the tip of a new root or shoot. Growth from any such meristem at the tip of a root or shoot is termed primary growth and results in the lengthening of that root or shoot. Secondary growth results in widening of a root or shoot from divisions of cells in a cambium.

In addition to growth by cell division, a plant may grow through cell elongation. This occurs when individual cells or groups of cells grow longer. Not all plant cells grow to the same length. When cells on one side of a stem grow longer and faster than cells on the other side, the stem bends to the side of the slower growing cells as a result. This directional growth can occur via a plant's response to a particular stimulus, such as light (phototropism), gravity (gravitropism), water, (hydrotropism), and physical contact (thigmotropism).

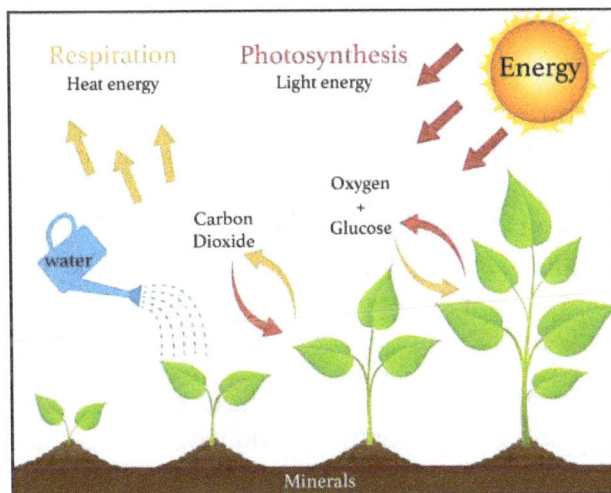

This image shows the development of a normal plant. It resembles the different growth processes for a leaf, a stem, etc. On top of the gradual growth of the plant, the image reveals the true meaning of phototropism and cell elongation, meaning the light energy from the sun is causing the growing plant to bend towards the light aka elongate.

Plant growth and development are mediated by specific plant hormones and plant growth regulators (PGRs). Endogenous hormone levels are influenced by plant age, cold hardiness, dormancy, and other metabolic conditions; photoperiod, drought, temperature, and other external environmental conditions; and exogenous sources of PGRs, e.g., externally applied and of rhizospheric origin.

Morphological Variation during Growth

Plants exhibit natural variation in their form and structure. While all organisms vary from individual to individual, plants exhibit an additional type of variation. Within a single individual, parts are repeated which may differ in form and structure from other similar parts. This variation is most easily seen in the leaves of a plant, though other organs such as stems and flowers may show similar variation. There are three primary causes of this variation: positional effects, environmental effects, and juvenility.

A sample of the variation in leaf shape that may occur on a single tree.

There is variation among the parts of a mature plant resulting from the relative position where the organ is produced. For example, along a new branch the leaves may vary in a consistent pattern along the branch. The form of leaves produced near the base of the branch differs from leaves produced at the tip of the plant, and this difference is consistent from branch to branch on a given plant and in a given species.

The way in which new structures mature as they are produced may be affected by the point in the plants life when they begin to develop, as well as by the environment to which the structures are exposed. Temperature has a multiplicity of effects on plants depending on a variety of factors, including the size and condition of the plant and the temperature and duration of exposure. The smaller and more succulent the plant, the greater the susceptibility to damage or death from temperatures that are too high or too low. Temperature affects the rate of biochemical and physiological processes, rates generally (within limits) increasing with temperature.

Juvenility or heteroblasty is when the organs and tissues produced by a young plant, such as a seedling, are often different from those that are produced by the same plant when it is older. For example, young trees will produce longer, leaner branches that grow upwards more than the branches they will produce as a fully grown tree. In addition, leaves produced during early growth tend to be larger, thinner, and more irregular than leaves on the adult plant. Specimens of juvenile plants may look so completely different from adult plants of the same species that egg-laying insects do not recognize the plant as food for their young. The transition from early to late growth forms is referred to as 'vegetative phase change'.

Adventitious Structures

Plant structures, including, roots, buds, and shoots, that develop in unusual locations are called adventitious. Such structures are common in vascular plants.

Adventitious roots and buds usually develop near the existing vascular tissues so they can connect to the xylem and phloem. However, the exact location varies greatly. In young stems, adventitious roots often form from parenchyma between the vascular bundles. In stems with secondary growth, adventitious roots often originate in phloem parenchyma near the vascular cambium. In stem cuttings, adventitious roots sometimes also originate in the callus cells that form at the cut surface. Leaf cuttings of the Crassula form adventitious roots in the epidermis.

Buds and Shoots

Adventitious buds develop from places other than a shoot apical meristem, which occurs at the tip of a stem, or on a shoot node, at the leaf axil, the bud being left there during the primary growth. They may develop on roots or leaves, or on shoots as a new growth. Shoot apical meristems produce one or more axillary or lateral buds at each node. When stems produce considerable secondary growth, the axillary buds may be destroyed. Adventitious buds may then develop on stems with secondary growth.

Adventitious buds are often formed after the stem is wounded or pruned. The adventitious buds help to replace lost branches. Adventitious buds and shoots also may develop on mature tree trunks when a shaded trunk is exposed to bright sunlight because surrounding trees are cut down. Redwood (Sequoia sempervirens) trees often develop many adventitious buds on their lower trunks. If the main trunk dies, a new one often sprouts from one of the adventitious buds. Small pieces of redwood trunk are sold as souvenirs termed redwood burls. They are placed in a pan of water, and the adventitious buds sprout to form shoots.

Some plants normally develop adventitious buds on their roots, which can extend quite a distance from the plant. Shoots that develop from adventitious buds on roots are termed suckers. They are a type of natural vegetative reproduction in many species, e.g. many grasses, quaking aspen and Canada thistle. The Pando quaking aspen grew from one trunk to 47,000 trunks via adventitious bud formation on a single root system.

Some leaves develop adventitious buds, which then form adventitious roots, as part of vegetative reproduction; e.g. piggyback plant (Tolmiea menziesii) and mother-of-thousands (Kalanchoe daigremontiana). The adventitious plantlets then drop off the parent plant and develop as separate clones of the parent.

Coppicing is the practice of cutting tree stems to the ground to promote rapid growth of adventitious shoots. It is traditionally used to produce poles, fence material or firewood. It is also practiced for biomass crops grown for fuel, such as poplar or willow.

Roots

Adventitious rooting may be a stress-avoidance acclimation for some species, driven by such inputs as hypoxia or nutrient deficiency. Another ecologically important function of adventitious rooting is the vegetative reproduction of tree species such as Salix and Sequoia in riparian settings.

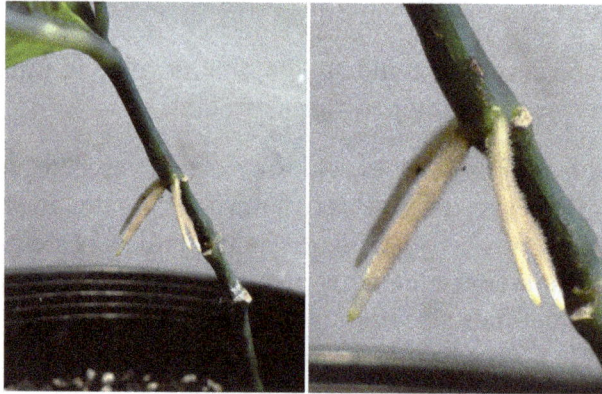

Roots forming above ground on a
cutting of Odontonema aka Firespike.

The ability of plant stems to form adventitious roots is utilised in commercial propagation by cuttings. Understanding of the physiological mechanisms behind adventitious rooting has allowed some progress to be made in improving the rooting of cuttings by the application of synthetic auxins as rooting powders and by the use of selective basal wounding. Further progress can be made in future years by applying research into other regulatory mechanisms to commercial propagation and by the comparative analysis of molecular and ecophysiological control of adventitious rooting in 'hard to root' vs. 'easy to root' species.

Adventitious roots and buds are very important when people propagate plants via cuttings, layering, tissue culture. Plant hormones, termed auxins, are often applied to stem, shoot or leaf cuttings to promote adventitious root formation, e.g. African violet and sedum leaves and shoots of poinsettia and coleus. Propagation via root cuttings requires adventitious bud formation, e.g. in horseradish and apple. In layering, adventitious roots are formed on aerial stems before the stem section is removed to make a new plant. Large houseplants are often propagated by air layering. Adventitious roots and buds must develop in tissue culture propagation of plants.

Modified Forms

- Tuberous roots lack a definite shape; example: sweet potato.

- Fasciculated root (tuberous root) occur in clusters at the base of the stem; examples: asparagus, dahlia.

- Nodulose roots become swollen near the tips; example: turmeric.

- Stilt roots arise from the first few nodes of the stem. These penetrate obliquely down into the soil and give support to the plant; examples: maize, sugarcane.

- Prop roots give mechanical support to aerial branches. The lateral branches grow vertically downward into the soil and act as pillars; example: banyan.

- Climbing roots arising from nodes attach themselves to some support and climb over it; example: money plant.

- Moniliform or beaded roots the fleshy roots give a beaded appearance, e.g.: bitter gourd, Portulaca, some grasses.

Leaf Development

The genetics behind leaf shape development in Arabidopsis thaliana has been broken down into three stages: The initiation of the leaf primordium, the establishment of dorsiventrality, and the development of a marginal meristem. Leaf primordium is initiated by the suppression of the genes and proteins of the class I KNOX family (such as SHOOT APICAL MERISTEM-LESS). These class I KNOX proteins directly suppress gibberellin biosynthesis in the leaf primodium. Many genetic factors were found to be involved in the suppression of these genes in leaf primordia (such as ASYMMETRIC LEAVES1, BLADE-ON-PETIOLE1, SAWTOOTH1, etc.). Thus, with this suppression, the levels of gibberellin increase and leaf primorium initiates growth.

Flower Development

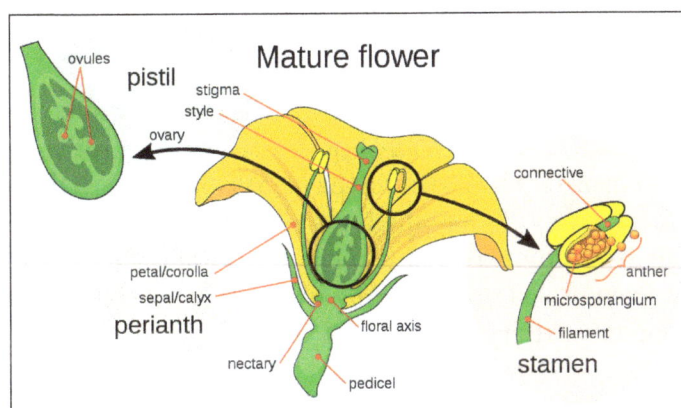

Anatomy of the flower.

Flower development is the process by which angiosperms produce a pattern of gene expression in meristems that leads to the appearance of an organ oriented towards sexual reproduction, the flower. There are three physiological developments that must occur in order for this to take place: firstly, the plant must pass from sexual immaturity into a sexually mature state (i.e. a transition towards flowering); secondly, the transformation of the apical meristem's function from a vegetative meristem into a floral meristem or inflorescence; and finally the growth of the flower's individual organs. The latter phase has been modelled using the ABC model, which describes the biological basis of the process from the perspective of molecular and developmental genetics.

An external stimulus is required in order to trigger the differentiation of the meristem into a flower meristem. This stimulus will activate mitotic cell division in the meristem, particularly on its sides where new primordia are formed. This same stimulus will also cause the meristem to follow a developmental pattern that will lead to the growth of floral meristems as opposed to vegetative meristems. The main difference between these two types of meristem, apart from the obvious disparity between the objective organ, is the verticillate (or whorled) phyllotaxis, that is, the absence of stem elongation among the successive whorls or verticils of the primordium. These verticils follow an acropetal development, giving rise to sepals, petals, stamens and carpels. Another difference from vegetative axillary meristems is that the floral meristem is "determined", which means that, once differentiated, its cells will no longer divide.

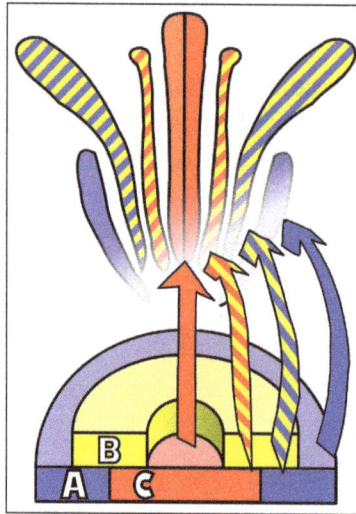

A diagram illustrating flower development in Arabidopsis.

The identity of the organs present in the four floral verticils is a consequence of the interaction of at least three types of gene products, each with distinct functions. According to the ABC model, functions A and C are required in order to determine the identity of the verticils of the perianth and the reproductive verticils, respectively. These functions are exclusive and the absence of one of them means that the other will determine the identity of all the floral verticils. The B function allows the differentiation of petals from sepals in the secondary verticil, as well as the differentiation of the stamen from the carpel on the tertiary verticil.

Floral Fragrance

Plants use floral form, flower, and scent to attract different insects for pollination. Certain compounds within the emitted scent appeal to particular pollinators. In *Petunia hybrida*, volatile benzenoids are produced to give off the floral smell. While components of the benzenoid biosynthetic pathway are known, the enzymes within the pathway, and subsequent regulation of those enzymes, are yet to be discovered.

To determine pathway regulation, P. hybrida Mitchell flowers were used in a petal-specific microarray to compare the flowers that were just about to produce the scent, to the P. hybrida cultivar W138 flowers that produce few volatile benzenoids. cDNAs of genes of both plants were sequenced. The results demonstrated that there is a transcription factor upregulated in the Mitchell flowers, but not in the W138 flowers lacking the floral aroma. This gene was named ODORANT1 (ODO1). To determine expression of ODO1 throughout the day, RNA gel blot analysis was done. The gel showed that ODO1 transcript levels began increasing between 1300 and 1600 h, peaked at 2200 h and were lowest at 1000 h. These ODO1 transcript levels directly correspond to the timeline of volatile benzenoid emission. Additionally, the gel supported the previous finding that W138 non-fragrant flowers have only one-tenth the ODO1 transcript levels of the Mitchell flowers. Thus, the amount of ODO1 made corresponds to the amount of volatile benzenoid emitted, indicating that ODO1 regulates benzenoid biosynthesis.

Additional genes contributing to the biosynthesis of major scent compounds are OOMT1 and OOMT2. OOMT1 and OOMT2 help to synthesize orcinol O-methyltransferases (OOMT), which catalyze the last two steps of the DMT pathway, creating 3,5-dimethoxytoluene (DMT). DMT is a scent compound produced by many different roses yet, some rose varieties, like Rosa gallica and Damask rose Rosa damascene, do not emit DMT. It has been suggested that these varieties do not make DMT because they do not have the OOMT genes. However, following an immunolocalization experiment, OOMT was found in the petal epidermis. To study this further, rose petals were subjected to ultracentrifugation. Supernatants and pellets were inspected by western blot. Detection of OOMT protein at 150,000g in the supernatant and the pellet allowed for researchers to conclude that OOMT protein is tightly associated with petal epidermis membranes. Such experiments determined that OOMT genes do exist within Rosa gallica and Damask rose Rosa damascene varieties, but the OOMT genes are not expressed in the flower tissues where DMT is made.

Photomorphogenesis

Photomorphogenesis is light-mediated development, where plant growth patterns respond to the light spectrum. This is a completely separate process from photosynthesis where light is used as a source of energy. Phytochromes, cryptochromes, and phototropins are photochromic sensory receptors that restrict the photomorphogenic effect of light to the UV-A, UV-B, blue, and red portions of the electromagnetic spectrum.

The photomorphogenesis of plants is often studied by using tightly frequency-controlled light sources to grow the plants. There are at least three stages of plant development where photomorphogenesis occurs: seed germination, seedling development, and the switch from the vegetative to the flowering stage (photoperiodism).

Most research on photomorphogenesis comes from plants, it occurs in several kingdoms: Fungi, Monera, Protista, and Plantae.

Developmental Stages Affected

Seed Germination

Light has profound effects on the development of plants. The most striking effects of light are observed when a germinating seedling emerges from the soil and is exposed to light for the first time.

Normally the seedling radicle (root) emerges first from the seed, and the shoot appears as the root becomes established. Later, with growth of the shoot (particularly when it emerges into the light) there is increased secondary root formation and branching. In this coordinated progression of developmental responses are early manifestations of correlative growth phenomena where the root affects the growth of the shoot and vice versa. To a large degree, the growth responses are hormone mediated.

Seedling Development

In the absence of light, plants develop an etiolated growth pattern. Etiolation of the seedling causes it to become elongated, which may facilitate it emerging from the soil.

A seedling that emerges in darkness follows a developmental program known as skotomorphogenesis (dark development), which is characterized by etiolation. Upon exposure to light, the seedling switches rapidly to photomorphogenesis (light development).

There are differences when comparing dark-grown (etiolated) and light-grown (de-etiolated) seedlings.

A dicot seedling emerging from the ground displays an apical hook (in the hypocotyl in this case), a response to dark conditions.

Etiolated characteristics:

- Distinct apical hook (dicot) or coleoptile (monocot)
- No leaf growth
- No chlorophyll
- Rapid stem elongation
- Limited radial expansion of stem
- Limited root elongation
- Limited production of lateral roots

De-etiolated characteristics:

- Apical hook opens or coleoptile splits open
- Leaf growth promoted
- Chlorophyll produced
- Stem elongation suppressed
- Radial expansion of stem
- Root elongation promoted
- Lateral root development accelerated

The developmental changes characteristic of photomorphogenesis shown by de-etiolated seedlings, are induced by light.

Photoperiodism

Some plants rely on light signals to determine when to switch from the vegetative to the flowering stage of plant development. This type of photomorphogenesis is known as photoperiodism and involves using red photoreceptors (phytochromes) to determine the daylength. As a result, photoperiodic plants only start making flowers when the days have reached a "critical daylength," allowing these plants to initiate their flowering period according to the time of year. For example, "long day" plants need long days to start flowering, and "short day" plants need to experience short days before they will start making flowers.

Photoperiodism also has an effect on vegetative growth, including on bud dormancy in perennial plants, though this is not as well-documented as the effect of photoperiodism on the switch to the flowering stage.

Light Receptors for Photomorphogenesis

Typically, plants are responsive to wavelengths of light in the blue, red and far-red regions of the spectrum through the action of several different photosensory systems. The photoreceptors for red and far-red wavelengths are known as phytochromes. There are at least 5 members of the phytochrome family of photoreceptors. There are several blue light photoreceptors known as cryptochromes. The combination of phytochromes and cryptochromes mediate growth and the flowering of plants in response to red light, far-red light, and blue light.

Red/Far-red Light

Plants use phytochrome to detect and respond to red and far-red wavelengths. Phytochromes are signaling proteins that promote photomorphogenesis in response to red light and far-red light. Phytochrome is the only known photoreceptor that absorbs light in the red/far red spectrum of light (600-750 nm) specifically and only for photosensory purposes. Phytochromes are proteins with a light absorbing pigment attached called a chromophore. The chromophore is a linear tetrapyrrole called phytochromobilin.

There are two forms of phytochromes: red light absorbing, Pr, and far-red light absorbing, Pfr. Pfr, which is the active form of phytochromes, can be reverted to Pr, which is the inactive form, slowly by inducing darkness or more rapidly by irradiation by far-red light. The phytochrome apoprotein, a protein that together with a prosthetic group forms a particular biochemical molecule such as a hormone or enzyme, is synthesized in the Pr form. Upon binding the chromophore, the holoprotein, an apoprotein combined with its prosthetic group, becomes sensitive to light. If it absorbs red light it will change conformation to the biologically active Pfr form. The Pfr form can absorb far red light and switch back to the Pr form. The Pfr promotes and regulates photomorphogenesis in response to FR light, whereas Pr regulates de-etiolation in response to R light.

Most plants have multiple phytochromes encoded by different genes. The different forms of phytochrome control different responses but there is also redundancy so that in the absence of one phytochrome, another may take on the missing functions. There are five genes that encode phytochromes in the Arabidopsis thaliana genetic model, PHYA-PHYE. PHYA is involved in the regulation of photomorphogenesis in response to far-red light. PHYB is involved in regulating photoreversible seed germination in response to red light. PHYC mediates the response between PHYA and PHYB. PHYD and PHYE mediate elongation of the internode and control the time in which the plant flowers.

Molecular analyses of phytochrome and phytochrome-like genes in higher plants (ferns, mosses, algae) and photosynthetic bacteria have shown that phytochromes evolved from prokaryotic photoreceptors that predated the origin of plants.

Takuma Tanada observed that the root tips of barley adhered to the sides of a beaker with a negatively charged surface after being treated with red light, yet released after being exposed to far-red light. For mung bean it was the opposite, where far-red light exposure caused the root tips to adhere, and red light caused the roots to detach. This effect of red and far-red light on root tips is now known as the Tanada effect.

Blue Light

Plants contain multiple blue light photoreceptors which have different functions. Based on studies with action spectra, mutants and molecular analyses, it has been determined that higher plants contain at least 4, and probably 5, different blue light photoreceptors.

Cryptochromes were the first blue light receptors to be isolated and characterized from any organism, and are responsible for the blue light reactions in photomorphogenesis. The proteins use a flavin as a chromophore. The cryptochromes have evolved from microbial DNA-photolyase, an enzyme that carries out light-dependent repair of UV damaged DNA. There are two different forms of cryptochromes that have been identified in plants, CRY1 and CRY2, which regulate the inhibition of hypocotyl elongation in response to blue light. Cryptochromes control stem elongation, leaf expansion, circadian rhythms and flowering time. In addition to blue light, cryptochromes also perceive long wavelength UV irradiation (UV-A). Since the cryptochromes were discovered in plants, several labs have identified homologous genes and photoreceptors in a number of other organisms, including humans, mice and flies.

There are blue light photoreceptors that are not a part of photomorphogenesis. For example, phototropin is the blue light photoreceptor that controls phototropism.

UV Light

Plants show various responses to UV light. UVR8 has been shown to be a UV-B receptor. Plants undergo distinct photomorphogenic changes as a result of UV-B radiation. They have photoreceptors that initiate morphogenetic changes in the plant embryo (hypocotyl, epicotyl, radicle) Exposure to UV- light in plants mediates biochemical pathways, photosynthesis, plant growth and many other processes essential to plant development. The UV-B photoreceptor, UV Resistance Locus8 (UVR8) detects UV-B rays and elicits photomorphogenic responses. These response are important for initiating hypocotyl elongation, leaf expansion, biosynthesis of flavonoids and many other important processes that affect the root-shoot system. Exposure to UV-B rays can be damaging to DNA inside of the plant cells, however, UVR8 induces genes required to acclimate plants to UV-B radiation, these genes are responsible for many biosynthesis pathways that involve protection from UV damage, oxidative stress, and photorepair of DNA damage.

There is still much to be discovered about the mechanisms involved in UV-B radiation and UVR8. Scientists are working to understand the pathways responsible for plant UV receptors response to solar radiation in natural environments.

Morphological Variation

Inflorescences emerging from protective coverings.

Plant morphology or phytomorphology is the study of the physical form and external structure of plants. This is usually considered distinct from plant anatomy, which is the study of the internal structure of plants, especially at the microscopic level. Plant morphology is useful in the visual identification of plants.

Scope

Asclepias syriaca showing complex morphology of the flowers.

Plant morphology "represents a study of the development, form, and structure of plants, and, by implication, an attempt to interpret these on the basis of similarity of plan and origin". There are four major areas of investigation in plant morphology, and each overlaps with another field of the biological sciences.

Looking up into the branch structure of a Pinus sylvestris tree.

First of all, morphology is comparative, meaning that the morphologist examines structures in many different plants of the same or different species, then draws comparisons and formulates ideas about similarities. When structures in different species are believed to exist and develop as a result of common, inherited genetic pathways, those structures are termed homologous. For example, the leaves of pine, oak, and cabbage all look very different, but share certain basic structures and arrangement of parts. The homology of leaves is an easy conclusion to make. The plant morphologist goes further, and discovers that the spines of cactus also share the same basic structure and development as leaves in other plants, and therefore cactus spines are homologous to leaves as well. This aspect of plant morphology overlaps with the study of plant evolution and paleobotany.

Secondly, plant morphology observes both the vegetative (somatic) structures of plants, as well as the reproductive structures. The vegetative structures of vascular plants includes the study of the shoot system, composed of stems and leaves, as well as the root system. The reproductive structures are more varied, and are usually specific to a particular group of plants, such as flowers and seeds, fern sori, and moss capsules. The detailed study of reproductive structures in plants led to the discovery of the alternation of generations found in all plants and most algae. This area of plant morphology overlaps with the study of biodiversity and plant systematics.

Thirdly, plant morphology studies plant structure at a range of scales. At the smallest scales are ultrastructure, the general structural features of cells visible only with the aid of an electron microscope, and cytology, the study of cells using optical microscopy. At this scale, plant morphology overlaps with plant anatomy as a field of study. At the largest scale is the study of plant growth habit, the overall architecture of a plant. The pattern of branching in a tree will vary from species to species, as will the appearance of a plant as a tree, herb, or grass.

Fourthly, plant morphology examines the pattern of development, the process by which structures originate and mature as a plant grows. While animals produce all the body parts they will ever have from early in their life, plants constantly produce new tissues and structures throughout their life. A living plant always has embryonic tissues. The way in which new structures mature as they are produced may be affected by the point in the plant's life when they begin to develop, as well as by the environment to which the structures are exposed. A morphologist studies this process, the causes, and its result. This area of plant morphology overlaps with plant physiology and ecology.

A Comparative Science

A plant morphologist makes comparisons between structures in many different plants of the same or different species. Making such comparisons between similar structures in different plants tackles the question of why the structures are similar. It is quite likely that similar underlying causes of genetics, physiology, or response to the environment have led to this similarity in appearance. The result of scientific investigation into these causes can lead to one of two insights into the underlying biology:

- Homology: The structure is similar between the two species because of shared ancestry and common genetics.

- Convergence: The structure is similar between the two species because of independent adaptation to common environmental pressures.

Understanding which characteristics and structures belong to each type is an important part of understanding plant evolution. The evolutionary biologist relies on the plant morphologist to interpret structures, and in turn provides phylogenies of plant relationships that may lead to new morphological insights.

Homology

When structures in different species are believed to exist and develop as a result of common, inherited genetic pathways, those structures are termed homologous. For example, the leaves of pine, oak, and cabbage all look very different, but share certain basic structures and arrangement of parts. The homology of leaves is an easy conclusion to make. The plant morphologist goes further, and discovers that the spines of cactus also share the same basic structure and development as leaves in other plants, and therefore cactus spines are homologous to leaves as well.

Convergence

When structures in different species are believed to exist and develop as a result of common adaptive responses to environmental pressure, those structures are termed convergent. For example, the fronds of Bryopsis plumosa and stems of Asparagus setaceus both have the same feathery branching appearance, even though one is an alga and one is a flowering plant. The similarity in overall structure occurs independently as a result of convergence. The growth form of many cacti and species of Euphorbia is very similar, even though they belong to widely distant families. The similarity results from common solutions to the problem of surviving in a hot, dry environment.

Vegetative and Reproductive Characteristics

Plant morphology treats both the vegetative structures of plants, as well as the reproductive structures.

The vegetative (somatic) structures of vascular plants include two major organ systems: (1) a shoot system, composed of stems and leaves, and (2) a root system. These two systems are common to nearly all vascular plants, and provide a unifying theme for the study of plant morphology.

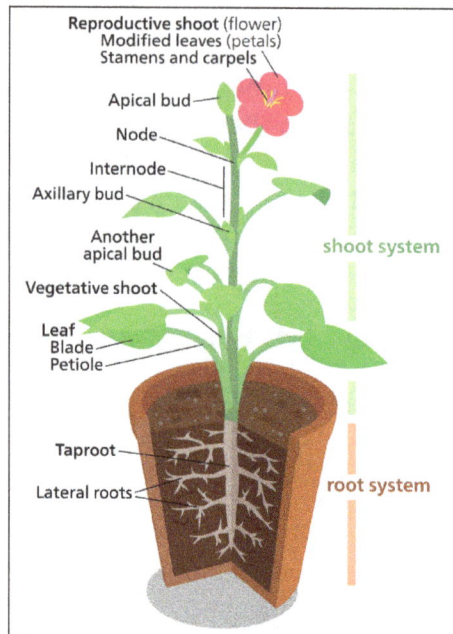

A diagram representing a "typical" eudicot.

By contrast, the reproductive structures are varied, and are usually specific to a particular group of plants. Structures such as flowers and fruits are only found in the angiosperms; sori are only found in ferns; and seed cones are only found in conifers and other gymnosperms. Reproductive characters are therefore regarded as more useful for the classification of plants than vegetative characters.

Use in Identification

Plant biologists use morphological characters of plants which can be compared, measured, counted and described to assess the differences or similarities in plant taxa and use these characters for plant identification, classification and descriptions.

When characters are used in descriptions or for identification they are called diagnostic or key characters which can be either qualitative and quantitative.

- Quantitative characters are morphological features that can be counted or measured for example a plant species has flower petals 10–12 mm wide.

- Qualitative characters are morphological features such as leaf shape, flower color or pubescence.

Both kinds of characters can be very useful for the identification of plants.

Alternation of Generations

The detailed study of reproductive structures in plants led to the discovery of the alternation of generations, found in all plants and most algae, by the German botanist Wilhelm Hofmeister. This discovery is one of the most important made in all of plant morphology, since it provides a common basis for understanding the life cycle of all plants.

Pigmentation in Plants

The primary function of pigments in plants is photosynthesis, which uses the green pigment chlorophyll along with several red and yellow pigments that help to capture as much light energy as possible. Pigments are also an important factor in attracting insects to flowers to encourage pollination.

Plant pigments include a variety of different kinds of molecule, including porphyrins, carotenoids, anthocyanins and betalains. All biological pigments selectively absorb certain wavelengths of light while reflecting others. The light that is absorbed may be used by the plant to power chemical reactions, while the reflected wavelengths of light determine the color the pigment will appear to the eye.

Morphology in Development

Plant development is the process by which structures originate and mature as a plant grows. It is a subject studies in plant anatomy and plant physiology as well as plant morphology.

The process of development in plants is fundamentally different from that seen in vertebrate animals. When an animal embryo begins to develop, it will very early produce all of the body parts that it will ever have in its life. When the animal is born (or hatches from its egg), it has all its body parts and from that point will only grow larger and more mature. By contrast, plants constantly produce new tissues and structures throughout their life from meristems located at the tips of organs, or between mature tissues. Thus, a living plant always has embryonic tissues.

The properties of organization seen in a plant are emergent properties which are more than the sum of the individual parts. "The assembly of these tissues and functions into an integrated multicellular organism yields not only the characteristics of the separate parts and processes but also quite a new set of characteristics which would not have been predictable on the basis of examination of the separate parts." In other words, knowing everything about the molecules in a plant are not enough to predict characteristics of the cells; and knowing all the properties of the cells will not predict all the properties of a plant's structure.

Growth

A vascular plant begins from a single celled zygote, formed by fertilisation of an egg cell by a sperm cell. From that point, it begins to divide to form a plant embryo through the process of embryogenesis. As this happens, the resulting cells will organize so that one end becomes the first root, while the other end forms the tip of the shoot. In seed plants, the embryo will develop one or more "seed leaves" (cotyledons). By the end of embryogenesis, the young plant will have all the parts necessary to begin in its life.

Once the embryo germinates from its seed or parent plant, it begins to produce additional organs (leaves, stems, and roots) through the process of organogenesis. New roots grow from root meristems located at the tip of the root, and new stems and leaves grow from shoot meristems located at the tip of the shoot. Branching occurs when small clumps of cells left behind by the meristem, and which have not yet undergone cellular differentiation to form a specialized tissue, begin to grow as the tip of a new root or shoot. Growth from any such meristem at the tip of a root or shoot is termed primary growth and results in the lengthening of that root or shoot. Secondary growth results in widening of a root or shoot from divisions of cells in a cambium.

In addition to growth by cell division, a plant may grow through cell elongation. This occurs when individual cells or groups of cells grow longer. Not all plant cells will grow to the same length. When cells on one side of a stem grow longer and faster than cells on the other side, the stem will bend to the side of the slower growing cells as a result. This directional growth can occur via a plant's response to a particular stimulus, such as light (phototropism), gravity (gravitropism), water, (hydrotropism), and physical contact (thigmotropism).

Plant growth and development are mediated by specific plant hormones and plant growth regulators (PGRs). Endogenous hormone levels are influenced by plant age, cold hardiness, dormancy, and other metabolic conditions; photoperiod, drought, temperature, and other external environmental conditions; and exogenous sources of PGRs, e.g., externally applied and of rhizospheric origin.

Morphological Variation

Plants exhibit natural variation in their form and structure. While all organisms vary from individual to individual, plants exhibit an additional type of variation. Within a single individual, parts are repeated which may differ in form and structure from other similar parts. This variation is most easily seen in the leaves of a plant, though other organs such as stems and flowers may show similar variation. There are three primary causes of this variation: positional effects, environmental effects, and juvenility.

Evolution of Plant Morphology

Transcription factors and transcriptional regulatory networks play key roles in plant morphogenesis and their evolution. During plant landing, many novel transcription factor families emerged and are preferentially wired into the networks of multicellular development, reproduction, and organ development, contributing to more complex morphogenesis of land plants.

Positional Effects

Variation in leaves from the giant ragweed illustrating positional effects.
The lobed leaves come from the base of the plant, while the unlobed
leaves come from the top of the plant.

Although plants produce numerous copies of the same organ during their lives, not all copies of a particular organ will be identical. There is variation among the parts of a mature plant resulting from the relative position where the organ is produced. For example, along a new branch the leaves may vary in a consistent pattern along the branch. The form of leaves produced near the base of the branch will differ from leaves produced at the tip of the plant, and this difference is consistent from branch to branch on a given plant and in a given species. This difference persists after the leaves at both ends of the branch have matured, and is not the result of some leaves being younger than others.

Environmental Effects

The way in which new structures mature as they are produced may be affected by the point in the plants life when they begin to develop, as well as by the environment to which the structures are exposed. This can be seen in aquatic plants and emergent plants.

Temperature

Temperature has a multiplicity of effects on plants depending on a variety of factors, including the size and condition of the plant and the temperature and duration of exposure. The smaller and more succulent the plant, the greater the susceptibility to damage or death from temperatures that are too high or too low. Temperature affects the rate of biochemical and physiological processes, rates generally (within limits) increasing with temperature. However, the Van't Hoff relationship for monomolecular reactions (which states that the velocity of a reaction is doubled or trebled by a temperature increase of 10 °C) does not strictly hold for biological processes, especially at low and high temperatures.

When water freezes in plants, the consequences for the plant depend very much on whether the freezing occurs intracellularly (within cells) or outside cells in intercellular (extracellular) spaces. Intracellular freezing usually kills the cell regardless of the hardiness of the plant and its tissues. Intracellular freezing seldom occurs in nature, but moderate rates of decrease in temperature, e.g., 1 °C to 6 °C/hour, cause intercellular ice to form, and this "extraorgan ice" may or may not be lethal, depending on the hardiness of the tissue.

At freezing temperatures, water in the intercellular spaces of plant tissues freezes first, though the water may remain unfrozen until temperatures fall below 7 °C. After the initial formation of ice intercellularly, the cells shrink as water is lost to the segregated ice. The cells undergo freeze-drying, the dehydration being the basic cause of freezing injury.

The rate of cooling has been shown to influence the frost resistance of tissues, but the actual rate of freezing will depend not only on the cooling rate, but also on the degree of supercooling and the properties of the tissue. Sakai demonstrated ice segregation in shoot primordia of Alaskan white and black spruces when cooled slowly to 30 °C to -40 °C. These freeze-dehydrated buds survived immersion in liquid nitrogen when slowly rewarmed. Floral primordia responded similarly. Extraorgan freezing in the primordia accounts for the ability of the hardiest of the boreal conifers to survive winters in regions when air temperatures often fall to -50 °C or lower. The hardiness of the winter buds of such conifers is enhanced by the smallness of the buds, by the evolution of faster translocation of water, and an ability to tolerate intensive freeze dehydration. In boreal species of Picea and Pinus, the frost resistance of 1-year-old seedlings is on a par with mature plants, given similar states of dormancy.

Juvenility

Juvenility in a seedling of European beech. There is a marked difference in shape between the first dark green "seed leaves" and the lighter second pair of leaves.

The organs and tissues produced by a young plant, such as a seedling, are often different from those that are produced by the same plant when it is older. This phenomenon is known as juvenility or heteroblasty. For example, young trees will produce longer, leaner branches that grow upwards more than the branches they will produce as a fully grown tree. In addition, leaves produced during early growth tend to be larger, thinner, and more irregular than leaves on the adult plant. Specimens of juvenile plants may look so completely different from adult plants of the same species that egg-laying insects do not recognize the plant as food for their young. Differences are seen in rootability and flowering and can be seen in the same mature tree. Juvenile cuttings taken from the base of a tree will form roots much more readily than cuttings originating from the mid to upper crown. Flowering close to the base of a tree is absent or less profuse than flowering in the higher branches especially when a young tree first reaches flowering age.

Plant Evolutionary Developmental Biology

Evolutionary developmental biology (evo-devo) is the study of developmental programs and patterns from an evolutionary perspective. It seeks to understand the various influences shaping the form and nature of life on the planet. Evo-devo arose as a separate branch of science rather recently. An early sign of this occurred in 1999.

Most of the synthesis in evo-devo has been in the field of animal evolution, one reason being the presence of elegant model systems like Drosophila melanogaster, C. elegans, zebrafish and Xenopus laevis. However, since 1980, a wealth of information on plant morphology, coupled with modern molecular techniques has helped shed light on the conserved and unique developmental patterns in the plant kingdom also.

Organisms, Databases and Tools

The most important model systems in plant development have been arabidopsis and maize. Maize has traditionally been the favorite of plant geneticists, while extensive resources in almost every area of plant physiology and development are available for Arabidopsis thaliana. Apart from these, rice, Antirrhinum majus, Brassica, and tomato are also being used in a variety of studies. The genomes of Arabidopsis thaliana and rice have been completely sequenced, while the others are in process. It must be emphasized here that the information from these "model" organisms form the basis of our developmental knowledge. While Brassica has been used primarily because of its convenient location in the phylogenetic tree in the mustard family, Antirrhinum majus is a convenient system for studying leaf architecture. Rice has been traditionally used for studying responses to hormones like abscissic acid and gibberelin as well as responses to stress. However, recently, not just the domesticated rice strain, but also the wild strains have been studied for their underlying genetic architectures.

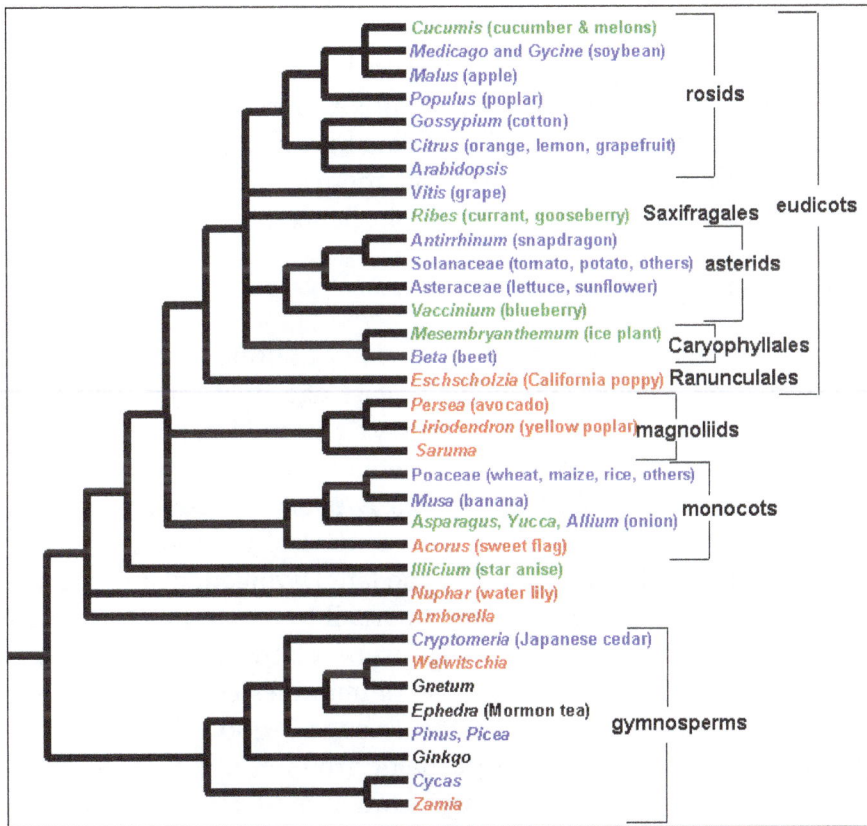

The sampling of the Floral Genome Project.

Some people have objected against extending the results of model organisms to the plant world. One argument is that the effect of gene knockouts in lab conditions wouldn't truly reflect even the same plant's response in the natural world. Also, these supposedly *crucial* genes might not be responsible for the evolutionary origin of that character. For these reasons, a comparative study of plant traits has been proposed as the way to go now.

Since the past few years, researchers have indeed begun looking at non-model, "non-conventional" organisms using modern genetic tools. One example of this is the Floral Genome Project, which envisages to study the evolution of the current patterns in the genetic architecture of the flower through comparative genetic analyses, with a focus on EST sequences. Like the FGP, there are several such ongoing projects that aim to find out conserved and diverse patterns in evolution of the plant shape. Expressed sequence tag (EST) sequences of quite a few non-model plants like sugarcane, apple, lotus, barley, cycas, coffee, to name a few, are available freely online. The Cycad Genomics Project, for example, aims to understand the differences in structure and function of genes between gymnosperms and angiosperms through sampling in the order Cycadales. In the process, it intends to make available information for the study of evolution of seeds, cones and evolution of life cycle patterns. Presently the most important sequenced genomes from an evo-devo point of view include those of A. thaliana (a flowering plant), poplar (a woody plant), Physcomitrella patens (a bryophyte), Maize (extensive genetic information), and Chlamydomonas reinhardtii (a green alga). The impact of such a vast amount of information on understanding common underlying developmental mechanisms can easily be realised.

Apart from EST and genome sequences, several other tools like PCR, yeast two-hybrid system, microarrays, RNA Interference, SAGE, QTL mapping etc. permit the rapid study of plant developmental patterns. Recently, cross-species hybridization has begun to be employed on microarray chips, to study the conservation and divergence in mRNA expression patterns between closely related species. Techniques for analyzing this kind of data have also progressed over the past decade. We now have better models for molecular evolution, more refined analysis algorithms and better computing power as a result of advances in computer sciences.

Evolution of Plant Morphology

Evidence suggests that an algal scum formed on the land 1,200 million years ago, but it was not until the Ordovician period, around 500 million years ago, that land plants appeared. These began to diversify in the late Silurian period, around 420 million years ago, and the fruits of their diversification are displayed in remarkable detail in an early Devonian fossil assemblage known as the Rhynie chert. This chert preserved early plants in cellular detail, petrified in volcanic springs. By the middle of the Devonian period most of the features recognised in plants today are present, including roots and leaves. By the late Devonian, plants had reached a degree of sophistication that allowed them to form forests of tall trees. Evolutionary innovation continued after the Devonian period. Most plant groups were relatively unscathed by the Permo-Triassic extinction event, although the structures of communities changed. This may have set the scene for the evolution of flowering plants in the Triassic (~200 million years ago), which exploded the Cretaceous and Tertiary. The latest major group of plants to evolve were the grasses, which became important in the mid Tertiary, from around 40 million years ago. The grasses, as well as many other groups, evolved new mechanisms of metabolism to survive the low CO_2 and warm, dry conditions of the tropics over the last 10 million years.

Meristems

The meristem architectures differ between angiosperms, gymnosperms and pteridophytes. The gymnosperm vegetative meristem lacks organization into distinct tunica and corpus layers. They possess large cells called central mother cells. In angiosperms, the outermost layer of cells divides anticlinally to generate the new cells, while in gymnosperms, the plane of division in the meristem differs for different cells. However, the apical cells do contain organelles like large vacuoles and starch grains, like the angiosperm meristematic cells.

Pteridophytes, like fern, on the other hand, do not possess a multicellular apical meristem. They possess a tetrahedral apical cell, which goes on to form the plant body. Any somatic mutation in this cell can lead to hereditary transmission of that mutation. The earliest meristem-like organization is seen in an algal organism from group *Charales* that has a single dividing cell at the tip, much like the pteridophytes, yet simpler. One can thus see a clear pattern in evolution of the meristematic tissue, from pteridophytes to angiosperms: Pteridophytes, with a single meristematic cell; gymnosperms with a multicellular, but less defined organization; and finally, angiosperms, with the highest degree of organization.

Evolution of Plant Transcriptional Regulation

Transcription factors and transcriptional regulatory networks play key roles in plant development and stress responses, as well as their evolution. During plant landing, many novel transcription factor families emerged and are preferentially wired into the networks of multicellular development, reproduction, and organ development, contributing to more complex morphogenesis of land plants.

Evolution of Leaves

Origins of the Leaf

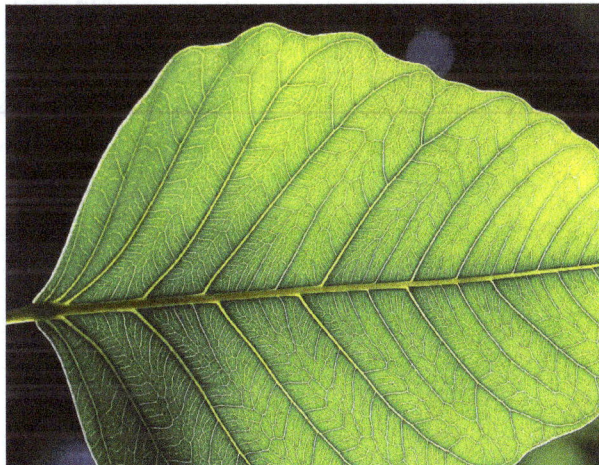

Leaf lamina: The leaf architecture probably
arose multiple times in the plant lineage.

Leaves are the primary photosynthetic organs of a plant. Based on their structure, they are classified into two types - microphylls, that lack complex venation patterns and megaphylls, that are large and with a complex venation. It has been proposed that these structures arose independently. Megaphylls, according to the telome theory, have evolved from plants that showed a three-dimensional branching architecture, through three transformations: planation, which involved formation of a planar architecture, webbing, or formation of the outgrowths between the planar branches and fusion, where these webbed outgrowths fused to form a proper leaf lamina. Studies have revealed that these three steps happened multiple times in the evolution of today's leaves.

Contrary to the telome theory, developmental studies of compound leaves have shown that, unlike simple leaves, compound leaves branch in three dimensions. Consequently, they appear partially homologous with shoots as postulated by Agnes Arber in her partial-shoot theory of the leaf. They appear to be part of a continuum between morphological categories, especially those of leaf and shoot. Molecular genetics confirmed these conclusions.

It has been proposed that the before the evolution of leaves, plants had the photosynthetic apparatus on the stems. Today's megaphyll leaves probably became commonplace some 360 mya, about 40 my after the simple leafless plants had colonized the land in the early Devonian period. This spread has been linked to the fall in the atmospheric carbon dioxide concentrations in the late Paleozoic era associated with a rise in density of stomata on leaf surface. This must have allowed for better transpiration rates and gas exchange. Large leaves with less stomata would have gotten heated up in the sun's heat, but an increased stomatal density allowed for a better-cooled leaf, thus making its spread feasible.

Factors Influencing Leaf Architectures

Various physical and physiological forces like light intensity, humidity, temperature, wind speeds etc. are thought to have influenced evolution of leaf shape and size. It is observed that high trees rarely have large leaves, owing to the obstruction they generate for winds. This obstruction can eventually lead to the tearing of leaves, if they are large. Similarly, trees that grow in temperate or taiga regions have pointed leaves, presumably to prevent nucleation of ice onto the leaf surface and reduce water loss due to transpiration. Herbivory, not only by large mammals, but also small insects has been implicated as a driving force in leaf evolution, an example being plants of the genus Aciphylla, that are commonly found in New Zealand. The now-extinct moas (birds) fed upon these plants, and the spines on the leaves probably discouraged the moas from feeding on them. Other members of Aciphylla that did not co-exist with the moas were spineless.

Genetic Evidences for Leaf Evolution

At the genetic level, developmental studies have shown that repression of the KNOX genes is required for initiation of the leaf primordium. This is brought about by *ARP* genes, which encode transcription factors. Genes of this type have been found in many plants studied till now, and the mechanism i.e. repression of KNOX genes in leaf primordia, seems to be quite conserved. Expression of KNOX genes in leaves produces complex leaves. It is speculated that the *ARP* function arose quite early in vascular plant evolution, because members of the primitive group lycophytes also have a functionally similar gene Other players that have a conserved role in defining leaf primordia are the phytohormone auxin, gibberelin and cytokinin.

One feature of a plant is its phyllotaxy. The arrangement of leaves on the plant body is such that the plant can maximally harvest light under the given constraints, and hence, one might expect the trait to be genetically robust. However, it may not be so. In maize, a mutation in only one gene called abphyl (abnormal phyllotaxy) was enough to change the phyllotaxy of the leaves. It implies that sometimes, mutational tweaking of a single locus on the genome is enough to generate diversity. The abphyl gene was later on shown to encode a cytokinin response regulator protein.

Once the leaf primordial cells are established from the SAM cells, the new axes for leaf growth are defined, one important among them being the abaxial-adaxial (lower-upper surface) axes. The genes involved in defining this, and the other axes seem to be more or less conserved among higher plants. Proteins of the HD-ZIPIII family have been implicated in defining the adaxial identity. These proteins deviate some cells in the leaf primordium from the default abaxial state, and make them adaxial. It is believed that in early plants with leaves, the leaves just had one type of surface - the abaxial one. This is the underside of today's leaves. The definition of the adaxial identity occurred some 200 million years after the abaxial identity was established. One can thus imagine the early leaves as an intermediate stage in evolution of today's leaves, having just arisen from spiny stem-like outgrowths of their leafless ancestors, covered with stomata all over, and not optimized as much for light harvesting.

The diversity of leaves.

How the infinite variety of plant leaves is generated is a subject of intense research. Some common themes have emerged. One of the most significant is the involvement of KNOX genes in generating compound leaves, as in tomato. But this again is not universal. For example, pea uses a different mechanism for doing the same thing. Mutations in genes affecting leaf curvature can also change leaf form, by changing the leaf from flat, to a crinky shape, like the shape of cabbage leaves. There also exist different morphogen gradients in a developing leaf which define the leaf's axis. Changes in these morphogen gradients may also affect the leaf form. Another very important class of regulators of leaf development are the microRNAs, whose role in this process has just begun to be documented. The coming years should see a rapid development in comparative studies on leaf development, with many EST sequences involved in the process coming online.

Molecular genetics has also shed light on the relation between radial symmetry (characteristic of stems) and dorsiventral symmetry (typical for leaves). James (2009) stated that "it is now widely accepted that radiality (characteristic of most shoots) and dorsiventrality (characteristic of leaves) are but extremes of a continuous spectrum. In fact, it is simply the timing of the KNOX gene expression." In fact there is evidence for this continuum already at the beginning of land plant evolution. Furthermore, studies in molecular genetics confirmed that compound leaves are intermediate between simple leaves and shoots, that is, they are partially homologous with simple leaves and shoots, since "it is now generally accepted that compound leaves express both leaf and shoot properties". This conclusion was reached by several authors on purely morphological grounds.

Evolution of Flowers

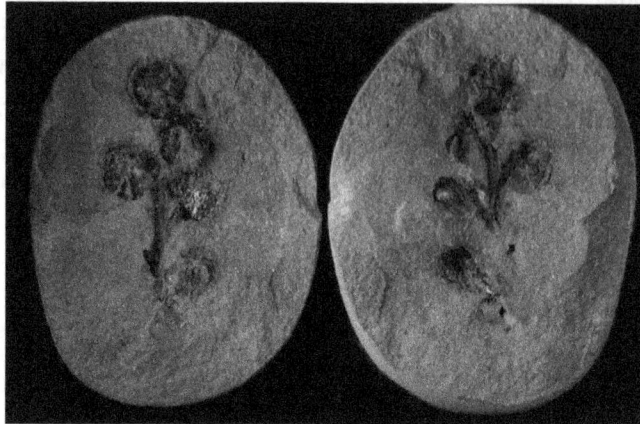

The pollen-bearing organs of the early flower Crossotheca.

Flower-like structures first appear in the fossil records some ~130 mya, in the Cretaceous era. The flowering plants have long been assumed to have evolved from within the gymnosperms; according to the traditional morphological view, they are closely allied to the gnetales. However, recent molecular evidence is at odds to this hypothesis, and further suggests that gnetales are more closely related to some gymnosperm groups than angiosperms, and that gymnosperms form a distinct clade to the angiosperms,. Molecular clock analysis predicts the divergence of flowering plants (anthophytes) and gymnosperms to ~300 mya.

The main function of a flower is reproduction, which, before the evolution of the flower and angiosperms, was the job of microsporophylls and megasporophylls. A flower can be considered a powerful evolutionary innovation, because its presence allowed the plant world to access new means and mechanisms for reproduction.

Origins of the Flower

It seems that on the level of the organ, the leaf may be the ancestor of the flower, or at least some floral organs. When we mutate some crucial genes involved in flower development, we end up with a cluster of leaf-like structures. Thus, sometime in history, the developmental program leading to formation of a leaf must have been altered to generate a flower. There probably also exists an overall robust framework within which the floral diversity has been generated. An example of that is a gene called LEAFY (LFY), which is involved in flower development in Arabidopsis thaliana. The homologs of this gene are found in angiosperms as diverse as tomato, snapdragon, pea, maize and even gymnosperms. Expression of Arabidopsis thaliana LFY in distant plants like poplar and citrus also results in flower-production in these plants. The LFY gene regulates the expression of some gene belonging to the MADS-box family. These genes, in turn, act as direct controllers of flower development.

Evolution of the MADS-box Family

The members of the MADS-box family of transcription factors play a very important and evolutionarily conserved role in flower development. According to the ABC model of flower development, three zones - A, B and C - are generated within the developing flower primordium, by the action of some transcription factors, that are members of the MADS-box family. Among these, the functions of the B and C domain genes have been evolutionarily more conserved than the A domain gene. Many of these genes have arisen through gene duplications of ancestral members of this family. Quite a few of them show redundant functions.

The evolution of the MADS-box family has been extensively studied. These genes are present even in pteridophytes, but the spread and diversity is many times higher in angiosperms. There appears to be quite a bit of pattern into how this family has evolved. Consider the evolution of the C-region gene AGAMOUS (AG). It is expressed in today's flowers in the stamens, and the carpel, which are reproductive organs. It's ancestor in gymnosperms also has the same expression pattern. Here, it is expressed in the strobili, an organ that produces pollens or ovules. Similarly, the B-genes' (AP3 and PI) ancestors are expressed only in the male organs in gymnosperms. Their descendants in the modern angiosperms also are expressed only in the stamens, the male reproductive organ. Thus, the same, then-existing components were used by the plants in a novel manner to generate the first flower. This is a recurring pattern in evolution.

Factors Influencing Floral Diversity

The various shapes and colors of flowers.

How is the enormous diversity in the shape, color and sizes of flowers established? There is enormous variation in the developmental program in different plants. For example, monocots possess structures like lodicules and palea, that were believed to be analogous to the dicot petals and carpels respectively.It turns out that this is true, and the variation is due to slight changes in the MADS-box genes and their expression pattern in the monocots. Another example is that of the toad-flax, Linaria vulgaris, which has two kinds of flower symmetries: radial and bilateral. These symmetries are due to epigenetic changes in just one gene called CYCLOIDEA.

Arabidopsis thaliana has a gene called AGAMOUS that plays an important role in defining how many petals and sepals and other organs are generated. Mutations in this gene give rise to the floral meristem obtaining an indeterminate fate, and many floral organs keep on getting produced. We have flowers like roses, carnations and morning glory, for example, that have very dense floral organs. These flowers have been selected by horticulturists since long for increased number of petals. Researchers have found that the morphology of these flowers is because of strong mutations in the

AGAMOUS homolog in these plants, which leads to them making a large number of petals and sepals. Several studies on diverse plants like petunia, tomato, impatiens, maize etc. have suggested that the enormous diversity of flowers is a result of small changes in genes controlling their development.

The large number of petals in roses has
probably been a result of human selection.

Some of these changes also cause changes in expression patterns of the developmental genes, resulting in different phenotypes. The Floral Genome Project looked at the EST data from various tissues of many flowering plants. The researchers confirmed that the ABC Model of flower development is not conserved across all angiosperms. Sometimes expression domains change, as in the case of many monocots, and also in some basal angiosperms like Amborella. Different models of flower development like the fading boundaries model, or the overlapping-boundaries model which propose non-rigid domains of expression, may explain these architectures. There is a possibility that from the basal to the modern angiosperms, the domains of floral architecture have gotten more and more fixed through evolution.

Flowering Time

Another floral feature that has been a subject of natural selection is flowering time. Some plants flower early in their life cycle, others require a period of vernalization before flowering. This decision is based on factors like temperature, light intensity, presence of pollinators and other environmental signals. In Arabidopsis thaliana it is known that genes like Constans (CO), Frigida, Flowering Locus C (FLC) and Flowering Locus T (FT) integrate the environmental signals and initiate the flower development pathway. Allelic variation in these loci have been associated with flowering time variations between plants. For example, Arabidopsis thaliana ecotypes that grow in the cold temperate regions require prolonged vernalization before they flower, while the tropical varieties and common lab strains, do not. Much of this variation is due to mutations in the FLC and Frigida genes, rendering them non-functional.

Many genes in the flowering time pathway are conserved across all plants studied to date. However, this does not mean that the mechanism of action is similarly conserved. For example, the monocot rice accelerates its flowering in short-day conditions, while Arabidopsis thaliana, a eudicot, responds to long-day conditions. In both plants, the proteins CO and FT are present but in Arabidopsis thaliana CO enhances FT production, while in rice the CO homolog represses FT production, resulting in completely opposite downstream effects.

Theories of Flower Evolution

There are many theories that propose how flowers evolved. Some of them are described below:

The Anthophyte Theory was based on the observation that a gymnospermic family Gnetaceae has a flower-like ovule. It has partially developed vessels as found in the angiosperms, and the megasporangium is covered by three envelopes, like the ovary structure of angiosperm flowers. However, many other lines of evidence show that gnetophytes are not related to angiosperms.

The Mostly Male Theory has a more genetic basis. Proponents of this theory point out that the gymnosperms have two very similar copies of the gene LFY while angiosperms only have one. Molecular clock analysis has shown that the other LFY paralog was lost in angiosperms around the same time as flower fossils become abundant, suggesting that this event might have led to floral evolution. According to this theory, loss of one of the LFY paralog led to flowers that were more male, with the ovules being expressed ectopically. These ovules initially performed the function of attracting pollinators, but sometime later, may have been integrated into the core flower.

Evolution of Secondary Metabolism

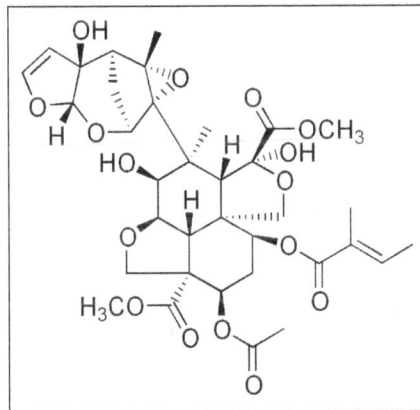

Structure of azadirachtin, a terpenoid produced by the neem plant, which helps ward off microbes and insects. Many secondary metabolites have complex structures.

Plant secondary metabolites are low molecular weight compounds, sometimes with complex structures that have no essential role in primary metabolism. They function in processes such as anti-herbivory, pollinator attraction, communication between plants, allelopathy, maintenance of symbiotic associations with soil flora and enhancing the rate of fertilization. Secondary metabolites have great structural and functional diversity and many thousands of enzymes may be involved in their synthesis, coded for by as much as 15–25% of the genome. Many plant secondary metabolites such as the colour and flavor components of saffron and the chemotherapeutic drug taxol are of culinary and medical significance to humans and are therefore of commercial importance.

Since bacteria possess the ability to make secondary metabolites such as the antibiotic penicillin, their production began quite early during evolution. But they assume more significant roles in eukaryotic and multicellular organisms. In plants they seem to have diversified using mechanisms such as gene duplications, evolution of novel genes and the development of novel biosynthetic pathways. Studies have shown that diversity in some of these compounds may be positively selected for. Cyanogenic glycosides may have been proposed to have evolved multiple times in different plant lineages, and there are several other instances of convergent evolution. For example, the enzymes for synthesis of limonene – a terpene – are more similar between angiosperms and gymnosperms than to their own terpene synthesis enzymes. This suggests independent evolution of the limonene biosynthetic pathway in these two lineages.

Mechanisms and Players in Evolution

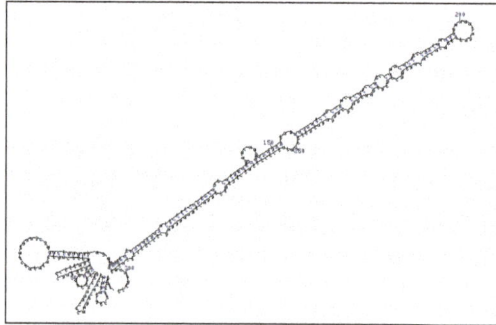

The stem-loop secondary structure of a
pri-microRNA from Brassica oleracea.

While environmental factors are significantly responsible for evolutionary change, they act merely as agents for natural selection. Some of the changes develop through interactions with pathogens. Change is inherently brought about via phenomena at the genetic level – mutations, chromosomal rearrangements and epigenetic changes. While the general types of mutations hold true across the living world, in plants, some other mechanisms have been implicated as highly significant.

Polyploidy is a very common feature in plants. It is believed that at least half plants are or have been polyploids. Polyploidy leads to genome doubling, thus generating functional redundancy in most genes. The duplicated genes may attain new function, either by changes in expression pattern or changes in activity. Polyploidy and gene duplication are believed to be among the most powerful forces in evolution of plant form. It is not known though, why genome doubling is such a frequent process in plants. One probable reason is the production of large amounts of secondary metabolites in plant cells. Some of them might interfere in the normal process of chromosomal segregation, leading to polypoidy.

Left: teosinte; right: maize;
middle: maize-teosinte hybrid.

In recent times, plants have been shown to possess significant microRNA families, which are conserved across many plant lineages. In comparison to animals, while the number of plant miRNA families is less, the size of each family is much larger. The miRNA genes are also much more spread out in the genome than those in animals, where they are found clustered. It has been proposed that these miRNA families have expanded by duplications of chromosomal regions. Many miRNA genes involved in regulation of plant development have been found to be quite conserved between plants studied.

Domestication of plants such as maize, rice, barley, wheat etc. has also been a significant driving force in their evolution. Some studies have looked at the origins of the maize plant and found that maize is a domesticated derivative of a wild plant from Mexico called teosinte. Teosinte belongs to the genus *Zea*, just as maize, but bears very small inflorescence, 5–10 hard cobs, and a highly branched and spread-out stem.

Cauliflower: Brassica oleracea var. botrytis.

Crosses between a particular teosinte variety and maize yield fertile offspring that are intermediate in phenotype between maize and teosinte. QTL analysis has also revealed some loci that when mutated in maize yield a teosinte-like stem or teosinte-like cobs. Molecular clock analysis of these genes estimates their origins to some 9000 years ago, well in accordance with other records of maize domestication. It is believed that a small group of farmers must have selected some maize-like natural mutant of teosinte some 9000 years ago in Mexico, and subjected it to continuous selection to yield the maize plant as known today.

Another case is that of cauliflower. The edible cauliflower is a domesticated version of the wild plant Brassica oleracea, which does not possess the dense undifferentiated inflorescence, called the curd, that cauliflower possesses.

Cauliflower possesses a single mutation in a gene called CAL, controlling meristem differentiation into inflorescence. This causes the cells at the floral meristem to gain an undifferentiated identity, and instead of growing into a flower, they grow into a lump of undifferentiated cells. This mutation has been selected through domestication at least.

Plant EvoDevo: Novel Approaches for Diverse Model Systems

For many years, a main focus of plant evolutionary developmental biology was studying the expression and phylogenetic history of genes implicated in developmental pathways. This approach has been enormously successful in identifying potentially conserved gene regulatory circuits that underlie major pattern formation processes in plants. Importantly, hypotheses were often generated on how changes in these gene regulatory-circuits led to the evolution of different plant forms. However, for quite some time, experimental testing of many of these hypotheses proved difficult, simply because the adequate molecular biology toolkit was not available across many plant lineages. This situation has changed dramatically in recent years. The advent of next generation sequencing considerably facilitated sequencing genomes and transcriptomes of plants throughout the phylogeny. Virus-induced gene silencing and the establishment of transformation methods for non-model plants enabled direct testing of gene functions on a wide phylogenetic spectrum, and elaborate biophysical techniques are increasingly applied to analyze changes in protein function during evolution. Furthermore, bioinformatics as well as systems biology are used to integrate the available data into a more coherent understanding of a fundamental question in plant evolution: What are the molecular underpinnings for the origin of different plant forms? Among the many facets this question touches are the transition to land, the emergence of vascular plants, the origin of the seed and the origin and diversification of floral form.

Among the novel experimental approaches that can be applied in a variety of systems, many researches offer perspectives on methodologies for the study of the diversification of form and the divergence of species. In interspecies gene transfer (IGT), researchers describe how candidate genes from a donor species are added to the wildtype genome of a recipient species to test their causality in the divergence of the two species. In evolutionary transgenomics, researchers describe the transfer of whole genomic fragments between species to identify novel genes of large effect without prior commitment to candidate genes. researchers. discuss the importance of biophysical studies for understanding morphological evolution. They show that small changes in the amino acid sequence of floral developmental regulators can lead to drastically altered protein–protein interaction patterns that may in turn have contributed to the evolution of the flower. This result illustrates that a more integrated approach—using genetics, biophysics and phylogenetics—is necessary to understand the evolution of development.

Several study present emerging model systems for the study of plant evo–devo, from shoot evolution to changes in inflorescence and floral traits. Some highlight the use of the emerging model fern Ceratopteris richardii, in the sister lineage to seed plants, for the study of the evolution and development of shoots, and of the genetic regulation of shoot apical meristems. Among the non-flowering seed plants, conifers represent a close extant relative of flowering plants, making them especially interesting from an evolutionary developmental perspective. others. highlight the importance of studying conifers and suggest that next generation sequencing and improved transformation protocols will make them more accessible to evo–devo studies. Within angiosperms, Vandenbussche et al. argue that new "supermodels" are required to comprehensively study the evolution of gene function that would ideally be as amenable to genetic analyses as Arabidopsis. The authors suggest that petunia could be one of those supermodels and provide an extensive overview of the genetic resources available for this system. Cronk et al. describe the evolution of catkin inflorescences in Salicaceae (poplars and willow) illustrating how the morphological richness of the Salicaceae coupled with the rapidly expanding genomic resources make this, of all woody plant families, particularly promising for genome-enabled evolutionary developmental biology. Landis et al. present Saltugilia (Polemoniaceae) as a model for the study of flower size (corolla tube length), a trait central to pollination syndrome. They find two independent evolutionary transitions to long corollas, and a correlation of long corollas with an increase in jigsaw cell size

and number and with the up regulation of genes associated with cell wall formation and organization. Morioka et al. examine floral diversity in Zingiberales, where members of Cannaceae have a laminar style that plays an important role in pollination interactions. Expression and evolution of genes involved in adaxial/abaxial polarity reveal a complex evolutionary history and suggest that loss of expression lead to this novel feature in Canna. Pabón-Mora et al. investigate the genetic basis of the highly derived and fused morphology of Aristolochia fimbriata. Developmental and comparative gene expression data support that the fused perianth is derived from sepals, not petals. Their data also reveal that A-class genes in the classic ABCE model do not contribute to perianth identity in this system. This finding provided further evidence that Arabidopsis A-class orthologs rarely contribute to perianth identity in other taxa.

There is evidence for opposing forces in the evolution of developmental mechanisms: conservation and divergence of gene and protein function. On the one hand, Hirakawa and Bowman show evidence for the conservation of protein function of the CLE family peptide hormone Tracheary element Differentiation Inhibitory Factor (TDIF) in regulating procambial cell fate, an important aspect in the evolutionary transition to vascular plants. The study performed evolutionary and functional comparative analyses, using protein assays, among representatives of major lineages in vascular plants and concluded that TDIF was integrated into shoot xylem differentiation in the euphyllophyte lineage (ferns and seed plants), after the split from lycophytes. Vialette-Guiraud et al. also present evidence for a conserved gene regulatory circuit, this time during flower development: They show that a genetic module consisting of microRNA164 and NAM transcription factors is responsible for the fusion of carpel margins in eurosids. The authors further suggest that the same gene regulatory circuit could have contributed to the emergence of the closed carpel very early during angiosperm evolution, and might thus have been involved in the origin of one of the most important evolutionary novelties in angiosperms. On the other hand, Yu et al. show that the exon-intron structure of genes is labile and influences the evolution of flower development gene lineages, frequently following gene duplication and speciation events. Their work suggests, for example, that an unstable gene structure in the AGL6 lineage may have contributed to its functional diversification in the flowering plants and to its divergence from the SEP lineage, which went on to become the major mediator of angiosperm floral quartets.

References

- Growth-and-development, tissues, plant-biology, biology, study-guides: cliffsnotes.com, Retrieved 3 August, 2019

- Azcón-bieto; et al. (2000). Fundamentos de fisiología vegetal. Mcgraw-hill/interamericana de españa, sau. Isbn 84-486-0258-7

- Brand, u; hobe, m; simon, r (2001). "functional domains in plant shoot meristems". Bioessays. 23 (2): 134–41. Doi:10.1002/1521-1878(200102)23:2<134::aid-bies1020>3.0.co;2-3. Pmid 11169586. Review

- Michael a dirr; charles w heuser, jr. (2006). "2". The reference manual of woody plant propagation (second ed.). Varsity press inc. Pp. 26, 28, 29. Isbn 0942375092

- B soltis, d.e.; soltis, p.s.; zanis, m.j. (2002). "phylogeny of seed plants based on evidence from eight genes" (abstract). American journal of botany. 89 (10): 1670–81. Doi:10.3732/ajb.89.10.1670. Pmid 21665594. Retrieved 2008-04-08

- Plant-reproductive-system, science: britannica.com, Retrieved 23 June, 2019

- Goodman c, coughlin b (2000). "the evolution of evo-devo biology". Proc. Natl. Acad. Sci. Usa. 97 (9): 4424–4425. Doi:10.1073/pnas.97.9.4424. Pmc 18255. Pmid 10781035

Evolutionary Biology

5

- **Evolutionary Biology**
- **Natural Selection**
- **Common Descent**
- **Speciation**
- **Adaptation**
- **Heredity**
- **Genotype**
- **Genetic Variation**
- **Modern Evolutionary Synthesis**

The subfield of biology which deals with the study of the evolutionary processes that produce the diversity of life on Earth is referred to as evolutionary biology. It includes processes such as natural selection, common descent and speciation. This chapter closely examines these processes of evolution as well as adaptation, heredity, genotype and genetic variation.

Evolutionary biology is the subfield of biology that deals with the processes of change in populations of organisms, especially taxonomy, paleontology, ethology, population genetics, and ecology.

Evolutionary theory has been extended almost continually since the evolutionary synthesis (ES), but except for the much greater importance afforded genetic drift, the principal tenets of the ES have been strongly supported. Adaptations are attributable to the sorting of genetic variation by natural selection, which remains the only known cause of increase in fitness. Mutations are not adaptively directed, but as principal authors of the ES recognized, the material (structural) bases of biochemistry and development affect the variety of phenotypic variations that arise by mutation and recombination. Against this historical background, we analyze major propositions in the movement for an 'extended evolutionary synthesis'. 'Niche construction' is a new label for a wide variety of well-known phenomena, many of which have been extensively studied, but some aspects may have been understudied. There is no reason to consider it a neglected 'process' of evolution. The proposition that phenotypic plasticity may engender new adaptive phenotypes that are later genetically assimilated or accommodated is theoretically plausible; it may be most likely when the new phenotype is not truly novel, but is instead a slight extension of a reaction norm already shaped by natural selection in similar environments. However, evolution in new environments often compensates for maladaptive plastic phenotypic responses. The union of population genetic theory with mechanistic understanding of developmental processes enables more complete understanding by joining ultimate and proximate causation; but the latter does not replace or invalidate the former. Newly discovered molecular phenomena have been easily accommodated in the past by elaborating orthodox evolutionary theory, and it appears that the same holds today for phenomena such as epigenetic inheritance. In several of these areas, empirical evidence is needed to evaluate enthusiastic speculation. Evolutionary theory will continue to be extended, but there is no sign that it requires emendation.

The current framework of evolutionary theory grew out of the evolutionary synthesis (ES), or the modern synthesis, it. In any discussion of extending or revising current theory, some understanding of the history of the ES and the subsequent development of the subject will be useful. the history of biology, and of evolutionary biology in particular, is one of generally gradual, rather than paradigm-shaking, development that builds successively on previous accomplishments. For example, soon after the discovery and canonization of Mendel's 'laws' in the earliest twentieth century, the 'law' of independent assortment had to be modified to account for linkage. The 'gene' went from a particulate 'factor' to a trinity of recon, muton and cistron (unit of recombination or mutation or function), thence to a protein-signifying code, and recently to an increasingly ambiguous functional part of a genome. Nevertheless, genetics has not cast out the old to accommodate the revolutionary new. Quite the opposite: classical Mendelian segregation, meiosis, linkage mapping and mutation are still important foundations of today's immensely more complex genetics.

The same holds for the evolutionary theory that has developed since the late 1920s. The ES remains, mutatis mutandis, the core of modern evolutionary biology. The ES included both the formulation of population genetic theory by Fisher, Haldane and Wright, and the interpretation of variation within species and of diverse information in zoology, botany and palaeontology.

Since the 1930s and 1940s, there has been a steady incorporation of new information, ranging from phylogeny and field studies of natural selection to evolutionary genomics and the panoply of genetic phenomena that could not have been imagined in the 1940s or even the 1960s—information that has informed (and sometimes been predicted by) a steady expansion of theory. Modern evolutionary biology recognizes and studies transposable elements, exon shuffling and chimeric genes, gene duplication and gene families, whole-genome duplication, *de novo* genes, gene regulatory networks, intragenomic conflict, kin selection, multilevel selection, phenotypic plasticity, maternal effects, morphological integration, evolvability, coevolution and more—some of these being phenomena and concepts unknown or dimly perceived a few decades ago.

Almost all of this amplification of evolutionary biology has been built on the core concepts of the ES, which have held fast with only modest modification. The most important tenets of the ES, I think, are these:

- The basic process of biological evolution is a population-level, not an individual-level, process that entails change not of the individual organism, but of the frequency of heritable variations within populations, from generation to generation. Dobzhansky defined evolution as change of allele frequencies, but some organism-focused evolutionary biologists, such as Rensch, Simpson and Mayr, had a more comprehensive conception of evolution, including phenotypic evolution, speciation and differential proliferation of clades, while recognizing that phenotypic evolution and speciation occur by changes in allele frequencies. When Rensch wrote of 'Evolution above the species level' and Simpson wrote about 'Tempo and mode in evolution', they were not talking about allele frequencies—although they recognized this as the elementary, generation by generation process of change.

- Heredity is based on 'genes', now understood to be DNA or RNA. DNA sequences transmitted in eukaryotes' gametes are not affected by an individual organism's experiences. Cultural inheritance has long been recognized, but insofar as it affects biological evolution, it does so by affecting natural selection. Some authors prefer to limit the term 'inheritance' to genetic transmission. For the sake of using a common language in this discourse, I will use the term 'inclusive inheritance' to include several forms of non-genetic 'inheritance', recognizing that this terminology may be disputed.

- Inherited variation arises by individually infrequent mutations; they are random in that their phenotypic effects, if any, are not directed towards 'need'. 'Random' should always be qualified by 'with respect to'; randomness of mutation has never meant that all possible alterations are equally likely, or that all genes mutate at the same rate, or that rates of mutation are immune from environmental factors (such as radiation and mutagens). Claims of 'directed mutation' have been shown to be groundless. The great majority of mutations that affect fitness are deleterious. Likewise, the direct effects of novel environments are more often harmful than beneficial: that is why they engender natural selection and adaptive change. These facts imply that we should be sceptical of the view that organisms are so constructed as to have well integrated, functional responses to mutations or novel environments. Without question, organisms have diverse homeostatic properties that buffer fitness against many environmental or genetic destabilizing events; but the maintenance of function depends on stabilizing or purifying natural selection.

- The frequencies of hereditary variants are altered by mutation (very slightly), gene flow, genetic drift, and natural selection. Directional or positive natural selection is the only known cause of adaptive change. Natural selection is not an agent, but a name for a consistent (biased, non-random) difference in the production of offspring by different classes of reproducing entities. The entities that were the focus of the ES were mostly phenotypically different individual organisms, but they can also be genes (as already recognized by Fisher, Haldane and Wright), populations or species.

- Species of sexually reproducing organisms are reproductively isolated groups of populations that arise by evolutionary divergence of geographically isolated (allopatric) populations.

Species evolve gradually, so not all populations can be classified into discrete species. Non-allopatric speciation is now recognized as possible, although its frequency is unknown.

- Large phenotypic changes of the kind that distinguish higher taxa and occur over long periods of time evolve gradually, as Darwin proposed, i.e. by the cumulation of relatively small incremental changes.

It is important to recognize that in population genetics theory, 'mutation' means any new alteration of the hereditary material that is stably transmitted across generations. The discovery of the molecular basis of heredity after the ES led to a greatly amplified understanding of evolutionary process and history, but the core theory of population genetics remained intact. For example, the core theory does not specify whether a mutation is a single base pair substitution, an insertion of a transposable element in a regulatory sequence, a gene duplication or a doubling of the entire genome. The framework of population genetics has incorporated new kinds of mutations, such as transposable elements, as they have been discovered.

Natural selection commonly was, and often still is, thought of as stemming from the ecological environment, but the forgers of the ES were well aware that selection had a far broader basis. Fisher described the evolution by selection of sex ratio, selfing and outcrossing, and he provided a genetic interpretation of Darwin's idea of sexual selection; Wright (who influenced Dobzhansky, who influenced Mayr) emphasized epistasis for fitness, in which prevalent alleles at one locus affect the selective value of alleles at another locus. Schmalhausen described 'internal selection'; mutations can have environment-independent effects on the function of physiological and developmental processes, and in turn on viability and reproduction. A causal account of any instance of selection requires different kinds of data—molecular, behavioural, ecological or other—but showing the existence of selection on a gene locus or a trait requires only data on components of fitness, such as rates of survival, fecundity or mating success.

Thus, the broad concepts of mutation and natural selection lack material content, in the sense that empirical data are needed to describe real instances of evolution, by identifying the agents of selection and the molecular and developmental basis of phenotypic variants. The conception of causes of evolution embodied in the synthetic theory, i.e. allele frequency change, differs from the 'structuralist' view of the causes of differences in morphology, physiology or behaviour that are commonly envisioned by mechanistic developmental biologists, physiologists or neurobiologists. A 'structuralist' approach to biology is cast in terms of the physical and chemical features of organisms, such as cell types and organs, and a 'structuralist' explanation of a morphological difference among species would be expressed in terms of signalling cascades, gene regulation and assembly of proteins into features that distinguish cell types. A complete account of any evolutionary change in phenotype would combine the two kinds of information: population genetic processes (causes of allele frequency change) together with the specific agents of selection and the structural and developmental basis of the altered phenotype. Part of the great power of the population genetic theory of evolutionary change lies in its generalization across diverse kinds of mutations, selective causes and phenotypic structures.

The leaders of the ES affirmed Darwin's gradualist view of long-term evolution, and rejected the saltationism of Schindewolf, Goldschmidt and others who supposed that higher taxa evolve by macromutations. Still, what might qualify as a 'large' mutational change was and is difficult to specify. Certainly some species differences and polymorphisms map to single loci with discretely different effects; phenomena such as neoteny (e.g. paedomorphic salamanders) were recognized, and nobody seemed to worry that partially paedomorphic salamanders were unknown and functionally unlikely. The 'instantaneous' origin of reproductively isolated species by polyploidy was likewise well known, especially in plants—but the species produced by polyploidy closely resemble the parent species: they are not new higher taxa.

Natural Selection

Modern biology began in the nineteenth century with Charles Darwin's work on evolution by natural selection.

Natural selection is the differential survival and reproduction of individuals due to differences in phenotype. It is a key mechanism of evolution, the change in the heritable traits characteristic of a population over generations. Charles Darwin popularised the term "natural selection", contrasting it with artificial selection, which in his view is intentional, whereas natural selection is not.

Variation exists within all populations of organisms. This occurs partly because random mutations arise in the genome of an individual organism, and offspring can inherit such mutations. Throughout the lives of the individuals, their genomes interact with their environments to cause variations in traits. The environment of a genome includes the molecular biology in the cell, other cells, other individuals, populations, species, as well as the abiotic environment. Because individuals with certain variants of the trait tend to survive and reproduce more than individuals with other, less successful variants, the population evolves. Other factors affecting reproductive success include sexual selection (now often included in natural selection) and fecundity selection.

Natural selection acts on the phenotype, the characteristics of the organism which actually interact with the environment, but the genetic (heritable) basis of any phenotype that gives that phenotype a reproductive advantage may become more common in a population. Over time, this process can result in populations that specialise for particular ecological niches (microevolution) and may eventually result in speciation (the emergence of new species, macroevolution). In other words, natural selection is a key process in the evolution of a population.

Natural selection is a cornerstone of modern biology. The concept, published by Darwin and Alfred Russel Wallace in a joint presentation of papers in 1858, was elaborated in Darwin's influential 1859 book On the Origin of Species by Means of Natural Selection, or the Preservation of Favoured Races in the Struggle for Life. He described natural selection as analogous to artificial selection, a process by which animals and plants with traits considered desirable by human breeders are systematically favoured for reproduction. The concept of natural selection originally developed in the absence of a valid theory of heredity; at the time of Darwin's writing, science had yet to develop modern theories of genetics. The union of traditional Darwinian evolution with subsequent discoveries in classical genetics formed the modern synthesis of the mid-20th century. The addition of molecular genetics has led to evolutionary developmental biology, which explains evolution at the molecular level. While genotypes can slowly change by random genetic drift, natural selection remains the primary explanation for adaptive evolution.

Mechanism

Heritable Variation and Differential Reproduction

During the industrial revolution, pollution killed many lichens, leaving tree trunks dark. A dark (melanic) morph of the peppered moth largely replaced the formerly usual light morph (both shown here). Since the moths are subject to predation by birds hunting by sight, the colour change offers better camouflage against the changed background, suggesting natural selection at work.

Natural variation occurs among the individuals of any population of organisms. Some differences may improve an individual's chances of surviving and reproducing such that its lifetime reproductive rate is increased, which means that it leaves more offspring. If the traits that give these individuals a reproductive advantage are also heritable, that is, passed from parent to offspring, then there will be differential reproduction, that is, a slightly higher proportion of fast rabbits or efficient algae in the next generation. Even if the reproductive advantage is very slight, over many generations any advantageous heritable trait becomes dominant in the population. In this way the natural environment of an organism "selects for" traits that confer a reproductive advantage, causing evolutionary change, as Darwin described. This gives the appearance of purpose, but in natural selection there is no intentional choice. Artificial selection is purposive where natural selection is not, though biologists often use teleological language to describe it.

The peppered moth exists in both light and dark colours in Great Britain, but during the industrial revolution, many of the trees on which the moths rested became blackened by soot, giving the dark-coloured moths an advantage in hiding from predators. This gave dark-coloured moths a better chance of surviving to produce dark-coloured offspring, and in just fifty years from the first dark moth being caught, nearly all of the moths in industrial Manchester were dark. The balance was reversed by the effect of the Clean Air Act 1956, and the dark moths became rare again, demonstrating the influence of natural selection on peppered moth evolution. A recent study, using image analysis and avian vision models, shows that pale individuals more closely match lichen backgrounds than dark morphs and for the first time quantifies the camouflage of moths to predation risk.

Fitness

The concept of fitness is central to natural selection. In broad terms, individuals that are more "fit" have better potential for survival, as in the well-known phrase "survival of the fittest", but the precise meaning of the term is much more subtle. Modern evolutionary theory defines fitness not by how long an organism lives, but by how successful it is at reproducing. If an organism lives half as long as others of its species, but has twice as many offspring surviving to adulthood, its genes become more common in the adult population of the next generation. Though natural selection acts on individuals, the effects of chance mean that fitness can only really be defined "on average" for the individuals within a population. The fitness of a particular genotype corresponds to the average effect on all individuals with that genotype. A distinction must be made between the concept of "survival of the fittest" and "improvement in fitness". "Survival of the fittest" does not give an "improvement in fitness", it only represents the removal of the less fit variants from a population. A mathematical example of "survival of the fittest" is given by Haldane in his paper "The Cost of Natural Selection". Haldane called this process "substitution" or more commonly in biology, this is called "fixation". This is correctly described by the differential survival and reproduction of individuals due to differences in phenotype. On the other hand, "improvement in fitness" is not dependent on the differential survival and reproduction of individuals due to differences in phenotype, it is dependent on the absolute survival of the particular variant. The probability of a beneficial mutation occurring on some member of a population depends on the total number of replications of that variant. The mathematics of "improvement in fitness was described by Kleinman.

An empirical example of "improvement in fitness" is given by the Kishony Mega-plate experiment. In this experiment, "improvement in fitness" depends on the number of replications of the particular variant for a new variant to appear that is capable of growing in the next higher drug concentration region. Fixation or substitution is not required for this "improvement in fitness". On the other hand, "improvement in fitness" can occur in an environment where "survival of the fittest" is also acting. The classic Lenski "E. coli long-term evolution experiment" is an example of adaptation in a competitive environment, ("improvement in fitness" during "survival of the fittest"). The probability of a beneficial mutation occurring on some member of the lineage to give improved fitness is slowed by the competition. The variant which is a candidate for a beneficial mutation in this limited carrying capacity environment must first out-compete the "less fit" variants in order to accumulate the requisite number of replications for there to be a reasonable probability of that beneficial mutation occurring.

Competition

In biology, competition is an interaction between organisms in which the fitness of one is lowered by the presence of another. This may be because both rely on a limited supply of a resource such as food, water, or territory. Competition may be within or between species, and may be direct or indirect. Species less suited to compete should in theory either adapt or die out, since competition plays a powerful role in natural selection, but according to the "room to roam" theory it may be less important than expansion among larger clades.

Competition is modelled by r/K selection theory, which is based on Robert MacArthur and E. O. Wilson's work on island biogeography. In this theory, selective pressures drive evolution in one of two stereotyped directions: r- or K-selection. These terms, r and K, can be illustrated in a logistic model of population dynamics:

$$\frac{dN}{dt} = rN\left(1 - \frac{N}{K}\right)$$

where r is the growth rate of the population (N), and K is the carrying capacity of its local environmental setting. Typically, r-selected species exploit empty niches, and produce many offspring, each with a relatively low probability of surviving to adulthood. In contrast, K-selected species are strong competitors in crowded niches, and invest more heavily in much fewer offspring, each with a relatively high probability of surviving to adulthood.

Classification

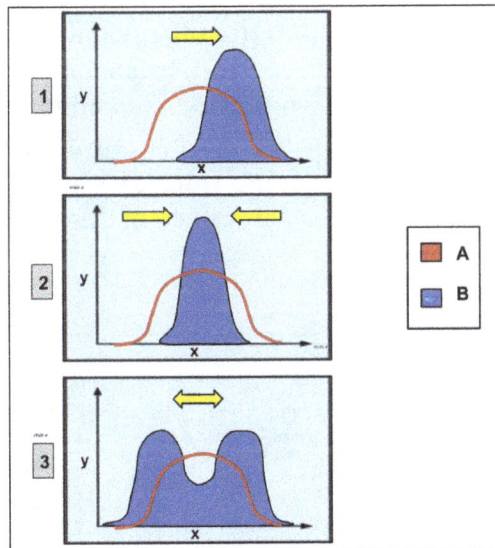

- Directional selection: A single extreme phenotype favoured.
- Stabilizing selection: Intermediate favoured over extremes.
- Disruptive selection: Extremes favoured over intermediate.
  - X-axis: Phenotypic trait
  - Y-axis: Number of organisms
  - Group A: Original population
  - Group B: After selection

Natural selection can act on any heritable phenotypic trait, and selective pressure can be produced by any aspect of the environment, including sexual selection and competition with members of the same or other species. However, this does not imply that natural selection is always directional and results in adaptive evolution; natural selection often results in the maintenance of the status quo by eliminating less fit variants.

Selection can be classified in several different ways, such as by its effect on a trait, on genetic diversity, by the life cycle stage where it acts, by the unit of selection, or by the resource being competed for.

By Effect on a Trait

Selection has different effects on traits. Stabilizing selection acts to hold a trait at a stable optimum, and in the simplest case all deviations from this optimum are selectively disadvantageous. Directional selection favours extreme values of a trait. The uncommon disruptive selection also acts during transition periods when the current mode is sub-optimal, but alters the trait in more than one direction. In particular, if the trait is quantitative and univariate then both higher and lower trait levels are favoured. Disruptive selection can be a precursor to speciation.

By Effect on Genetic Diversity

Alternatively, selection can be divided according to its effect on genetic diversity. Purifying or negative selection acts to remove genetic variation from the population (and is opposed by *de novo* mutation, which introduces new variation. In contrast, balancing selection acts to maintain genetic variation in a population, even in the absence of *de novo* mutation, by negative frequency-dependent selection. One mechanism for this is heterozygote advantage, where individuals with two different alleles have a selective advantage over individuals with just one allele. The polymorphism at the human ABO blood group locus has been explained in this way.

By Life Cycle Stage

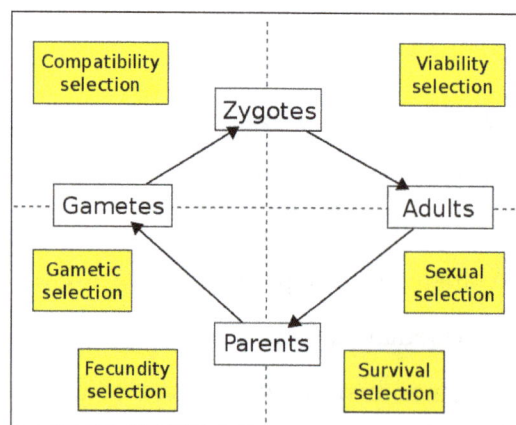

Different types of selection act at each life cycle stage of a sexually reproducing organism.

Another option is to classify selection by the life cycle stage at which it acts. Some biologists recognise just two types: viability (or survival) selection, which acts to increase an organism's probability of survival, and fecundity (or fertility or reproductive) selection, which acts to increase the rate of reproduction, given survival. Others split the life cycle into further components of selection. Thus viability and survival selection may be defined separately and respectively as acting to improve the probability of survival before and after reproductive age is reached, while fecundity selection may be split into additional sub-components including sexual selection, gametic selection, acting on gamete survival, and compatibility selection, acting on zygote formation.

By Unit of Selection

Selection can also be classified by the level or unit of selection. Individual selection acts on the individual, in the sense that adaptations are "for" the benefit of the individual, and result from selection among individuals. Gene selection acts directly at the level of the gene. In kin selection and intragenomic conflict, gene-level selection provides a more apt explanation of the underlying process. Group selection, if it occurs, acts on groups of organisms, on the assumption that groups replicate and mutate in an analogous way to genes and individuals. There is an ongoing debate over the degree to which group selection occurs in nature.

By Resource being Competed For

The peacock's elaborate plumage is mentioned by Darwin as an example of sexual selection, and is a classic example of Fisherian runaway, driven to its conspicuous size and coloration through mate choice by females over many generations.

Finally, selection can be classified according to the resource being competed for. Sexual selection results from competition for mates. Sexual selection typically proceeds via fecundity selection, sometimes at the expense of viability. Ecological selection is natural selection via any means other than sexual selection, such as kin selection, competition, and infanticide. Following Darwin, natural selection is sometimes defined as ecological selection, in which case sexual selection is considered a separate mechanism.

Sexual selection as first articulated by Darwin (using the example of the peacock's tail) refers specifically to competition for mates, which can be intrasexual, between individuals of the same sex, that is male–male competition, or intersexual, where one gender chooses mates, most often with males displaying and females choosing. However, in some species, mate choice is primarily by males, as in some fishes of the family Syngnathidae.

Phenotypic traits can be displayed in one sex and desired in the other sex, causing a positive feedback loop called a Fisherian runaway, for example, the extravagant plumage of some male birds such as the peacock. An alternate theory proposed by the same Ronald Fisher in 1930 is the sexy son hypothesis, that mothers want promiscuous sons to give them large numbers of grandchildren and so choose promiscuous fathers for their children. Aggression between members of the same sex is sometimes associated with very distinctive features, such as the antlers of stags, which are used in combat with other stags. More generally, intrasexual selection is often associated with sexual dimorphism, including differences in body size between males and females of a species.

Arms Races

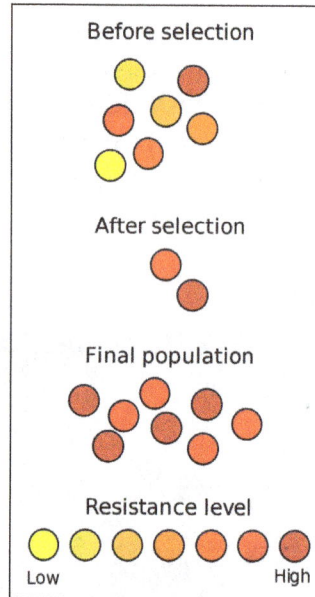

Selection in action: Resistance to antibiotics grows though the survival of individuals less affected by the antibiotic. Their offspring inherit the resistance.

Natural selection is seen in action in the development of antibiotic resistance in microorganisms. Since the discovery of penicillin in 1928, antibiotics have been used to fight bacterial diseases. The widespread misuse of antibiotics has selected for microbial resistance to antibiotics in clinical use, to the point that the methicillin-resistant Staphylococcus aureus (MRSA) has been described as a "superbug" because of the threat it poses to health and its relative invulnerability to existing drugs. Response strategies typically include the use of different, stronger antibiotics; however, new strains of MRSA have recently emerged that are resistant even to these drugs. This is an evolutionary arms race, in which bacteria develop strains less susceptible to antibiotics, while medical researchers attempt to develop new antibiotics that can kill them. A similar situation occurs with pesticide resistance in plants and insects. Arms races are not necessarily induced by man; a well-documented example involves the spread of a gene in the butterfly Hypolimnas bolina suppressing male-killing activity by Wolbachia bacteria parasites on the island of Samoa, where the spread of the gene is known to have occurred over a period of just five years.

Evolution by Means of Natural Selection

A prerequisite for natural selection to result in adaptive evolution, novel traits and speciation is the presence of heritable genetic variation that results in fitness differences. Genetic variation is the result of mutations, genetic recombinations and alterations in the karyotype (the number, shape, size and internal arrangement of the chromosomes). Any of these changes might have an effect that is highly advantageous or highly disadvantageous, but large effects are rare. In the past, most changes in the genetic material were considered neutral or close to neutral because they occurred in noncoding DNA or resulted in a synonymous substitution. However, many mutations in non-coding DNA have deleterious effects. Although both mutation rates and average fitness effects of mutations are dependent on the organism, a majority of mutations in humans are slightly deleterious.

Some mutations occur in "toolkit" or regulatory genes. Changes in these often have large effects on the phenotype of the individual because they regulate the function of many other genes. Most, but not all, mutations in regulatory genes result in non-viable embryos. Some nonlethal regulatory mutations occur in HOX genes in humans, which can result in a cervical rib or polydactyly, an increase in the number of fingers or toes. When such mutations result in a higher fitness, natural selection favours these phenotypes and the novel trait spreads in the population. Established traits are not immutable; traits that have high fitness in one environmental context may be much less fit if environmental conditions change. In the absence of natural selection to preserve such a trait, it becomes more variable and deteriorate over time, possibly resulting in a vestigial manifestation of the trait, also called evolutionary baggage. In many circumstances, the apparently vestigial structure may retain a limited functionality, or may be co-opted for other advantageous traits in a phenomenon known as preadaptation. A famous example of a vestigial structure, the eye of the blind mole-rat, is believed to retain function in photoperiod perception.

Speciation

Speciation requires a degree of reproductive isolation—that is, a reduction in gene flow. However, it is intrinsic to the concept of a species that hybrids are selected against, opposing the evolution of reproductive isolation, a problem that was recognised by Darwin. The problem does not occur in allopatric speciation with geographically separated populations, which can diverge with different sets of mutations. E. B. Poulton realized in 1903 that reproductive isolation could evolve through divergence, if each lineage acquired a different, incompatible allele of the same gene. Selection against the heterozygote would then directly create reproductive isolation, leading to the Bateson–Dobzhansky–Muller model, further elaborated by H. Allen Orr and Sergey Gavrilets. With reinforcement, however, natural selection can favor an increase in pre-zygotic isolation, influencing the process of speciation directly.

Genetic Basis

Genotype and Phenotype

Natural selection acts on an organism's phenotype, or physical characteristics. Phenotype is determined by an organism's genetic make-up (genotype) and the environment in which the organism lives. When different organisms in a population possess different versions of a gene for a certain trait, each of these versions is known as an allele. It is this genetic variation that underlies differences in phenotype. An example is the ABO blood type antigens in humans, where three alleles govern the phenotype.

Some traits are governed by only a single gene, but most traits are influenced by the interactions of many genes. A variation in one of the many genes that contributes to a trait may have only a small effect on the phenotype; together, these genes can produce a continuum of possible phenotypic values.

Directionality of Selection

When some component of a trait is heritable, selection alters the frequencies of the different alleles, or variants of the gene that produces the variants of the trait. Selection can be divided into three classes, on the basis of its effect on allele frequencies: directional, stabilizing, and purifying selection. Directional selection occurs when an allele has a greater fitness than others, so that it increases in frequency, gaining an increasing share in the population. This process can continue until the allele is fixed and the entire population shares the fitter phenotype. Far more common is stabilizing selection, which lowers the frequency of alleles that have a deleterious effect on the phenotype—that is, produce organisms of lower fitness. This process can continue until the allele is eliminated from the population. Purifying selection conserves functional genetic features, such as protein-coding genes or regulatory sequences, over time by selective pressure against deleterious variants.

Some forms of balancing selection do not result in fixation, but maintain an allele at intermediate frequencies in a population. This can occur in diploid species (with pairs of chromosomes) when heterozygous individuals (with just one copy of the allele) have a higher fitness than homozygous individuals (with two copies). This is called heterozygote advantage or over-dominance, of which the best-known example is the resistance to malaria in humans heterozygous for sickle-cell anaemia. Maintenance of allelic variation can also occur through disruptive or diversifying selection, which favours genotypes that depart from the average in either direction (that is, the opposite of over-dominance), and can result in a bimodal distribution of trait values. Finally, balancing selection can occur through frequency-dependent selection, where the fitness of one particular phenotype depends on the distribution of other phenotypes in the population. The principles of game theory have been applied to understand the fitness distributions in these situations, particularly in the study of kin selection and the evolution of reciprocal altruism.

Selection, Genetic Variation and Drift

A portion of all genetic variation is functionally neutral, producing no phenotypic effect or significant difference in fitness. Motoo Kimura's neutral theory of molecular evolution by genetic drift proposes that this variation accounts for a large fraction of observed genetic diversity. Neutral events can radically reduce genetic variation through population bottlenecks which among other things can cause the founder effect in initially small new populations. When genetic variation does not result in differences in fitness, selection cannot directly affect the frequency of such variation. As a result, the genetic variation at those sites is higher than at sites where variation does influence fitness. However, after a period with no new mutations, the genetic variation at these sites is eliminated due to genetic drift. Natural selection reduces genetic variation by eliminating maladapted individuals, and consequently the mutations that caused the maladaptation. At the same time, new mutations occur, resulting in a mutation–selection balance. The exact outcome of the two processes depends both on the rate at which new mutations occur and on the strength of the natural selection, which is a function of how unfavourable the mutation proves to be.

Genetic linkage occurs when the loci of two alleles are in close proximity on a chromosome. During the formation of gametes, recombination reshuffles the alleles. The chance that such a reshuffle occurs between two alleles is inversely related to the distance between them. Selective sweeps occur when an allele becomes more common in a population as a result of positive selection. As the prevalence of one allele increases, closely linked alleles can also become more common by "genetic hitchhiking", whether they are neutral or even slightly deleterious. A strong selective sweep results in a region of the genome where the positively selected haplotype (the allele and its neighbours) are in essence the only ones that exist in the population. Selective sweeps can be detected by measuring linkage disequilibrium, or whether a given haplotype is overrepresented in the population. Since a selective sweep also results in selection of neighbouring alleles, the presence of a block of strong linkage disequilibrium might indicate a 'recent' selective sweep near the centre of the block.

Background selection is the opposite of a selective sweep. If a specific site experiences strong and persistent purifying selection, linked variation tends to be weeded out along with it, producing a region in the genome of low overall variability. Because background selection is a result of deleterious new mutations, which can occur randomly in any haplotype, it does not produce clear blocks of linkage disequilibrium, although with low recombination it can still lead to slightly negative linkage disequilibrium overall.

Impact

Darwin's ideas, along with those of Adam Smith and Karl Marx, had a profound influence on 19th century thought, including his radical claim that "elaborately constructed forms, so different from each other, and dependent on each other in so complex a manner" evolved from the simplest forms of life by a few simple principles. This inspired some of Darwin's most ardent supporters—and provoked the strongest opposition. Natural selection had the power, according to Stephen Jay Gould, to "dethrone some of the deepest and most traditional comforts of Western thought", such as the belief that humans have a special place in the world.

In the words of the philosopher Daniel Dennett, "Darwin's dangerous idea" of evolution by natural selection is a "universal acid," which cannot be kept restricted to any vessel or container, as it soon leaks out, working its way into ever-wider surroundings. Thus, in the last decades, the concept of natural selection has spread from evolutionary biology to other disciplines, including evolutionary computation, quantum Darwinism, evolutionary economics, evolutionary epistemology, evolutionary psychology, and cosmological natural selection. This unlimited applicability has been called universal Darwinism.

Origin of Life

How life originated from inorganic matter remains an unresolved problem in biology. One prominent hypothesis is that life first appeared in the form of short self-replicating RNA polymers. On this view, life may have come into existence when RNA chains first experienced the basic conditions, as conceived by Charles Darwin, for natural selection to operate. These conditions are: heritability, variation of type, and competition for limited resources. The fitness of an early RNA replicator would likely have been a function of adaptive capacities that were intrinsic (i.e., determined by the nucleotide sequence) and the availability of resources. The three primary adaptive capacities could logically have been: (1) the capacity to replicate with moderate fidelity (giving rise to both heritability and variation of type), (2) the capacity to avoid decay, and (3) the capacity to acquire and process resources. These capacities would have been determined initially by the folded configurations (including those configurations with ribozyme activity) of the RNA replicators that, in turn, would have been encoded in their individual nucleotide sequences.

Cell and Molecular Biology

In 1881, the embryologist Wilhelm Roux published Der Kampf der Theile im Organismus (The Struggle of Parts in the Organism) in which he suggested that the development of an organism results from a Darwinian competition between the parts of the embryo, occurring at all levels, from molecules to organs. In recent years, a modern version of this theory has been proposed by Jean-Jacques Kupiec. According to this cellular Darwinism, random variation at the molecular level generates diversity in cell types whereas cell interactions impose a characteristic order on the developing embryo.

Social and Psychological Theory

The social implications of the theory of evolution by natural selection also became the source of continuing controversy. Friedrich Engels, a German political philosopher and co-originator of the ideology of communism, wrote in 1872 that "Darwin did not know what a bitter satire he wrote on mankind, and especially on his countrymen, when he showed that free competition, the struggle for existence, which the economists celebrate as the highest historical achievement, is the normal state of the animal kingdom." Herbert Spencer and the eugenics advocate Francis Galton's interpretation of natural selection as necessarily progressive, leading to supposed advances in intelligence and civilisation, became a justification for colonialism, eugenics, and social Darwinism. For example, in 1940, Konrad Lorenz, in writings that he subsequently disowned, used the theory as a justification for policies of the Nazi state. He wrote "selection for toughness, heroism, and social utility must be accomplished by some human institution, if mankind, in default of selective factors, is not to be ruined by domestication-induced degeneracy. The racial idea as the basis of our state has already accomplished much in this respect." Others have developed ideas that human societies and culture evolve by mechanisms analogous to those that apply to evolution of species.

More recently, work among anthropologists and psychologists has led to the development of sociobiology and later of evolutionary psychology, a field that attempts to explain features of human psychology in terms of adaptation to the ancestral environment. The most prominent example of evolutionary psychology, notably advanced in the early work of Noam Chomsky and later by Steven Pinker, is the hypothesis that the human brain has adapted to acquire the grammatical rules of natural language. Other aspects of human behaviour and social structures, from specific cultural norms such as incest avoidance to broader patterns such as gender roles, have been hypothesised to have similar origins as adaptations to the early environment in which modern humans evolved. By analogy to the action of natural selection on genes, the concept of memes—"units of cultural transmission," or culture's equivalents of genes undergoing selection and recombination—has arisen, first described in this form by Richard Dawkins in 1976 and subsequently expanded upon by philosophers such as Daniel Dennett as explanations for complex cultural activities, including human consciousness.

Information and Systems Theory

In 1922, Alfred J. Lotka proposed that natural selection might be understood as a physical principle that could be described in terms of the use of energy by a system, a concept later developed by Howard T. Odum as the maximum power principle in thermodynamics, whereby evolutionary systems with selective advantage maximise the rate of useful energy transformation.

The principles of natural selection have inspired a variety of computational techniques, such as "soft" artificial life, that simulate selective processes and can be highly efficient in 'adapting' entities to an environment defined by a specified fitness function. For example, a class of heuristic optimisation algorithms known as genetic algorithms, pioneered by John Henry Holland in the 1970s and expanded upon by David E. Goldberg, identify optimal solutions by simulated reproduction and mutation of a population of solutions defined by an initial probability distribution. Such algorithms are particularly useful when applied to problems whose energy landscape is very rough or has many local minima.

In Fiction

Darwinian evolution by natural selection is pervasive in literature, whether taken optimistically in terms of how humanity may evolve towards perfection, or pessimistically in terms of the dire consequences of the interaction of human nature and the struggle for survival. Among major responses is Samuel Butler's 1872 pessimistic *Erewhon* ("nowhere", written mostly backwards). In 1893 H. G. Wells imagined "The Man of the Year Million", transformed by natural selection into a being with a huge head and eyes, and shrunken body.

Common Descent

Common descent describes how, in evolutionary biology, a group of organisms share a most recent common ancestor. There is massive evidence of common descent of all life on Earth from the last universal common ancestor (LUCA). In July 2016, scientists reported identifying a set of 355 genes from the LUCA by comparing the genomes of the three domains of life, archaea, bacteria, and eukaryotes.

Common ancestry between organisms of different species arises during speciation, in which new species are established from a single ancestral population. Organisms which share a more-recent common ancestor are more closely related. The most recent common ancestor of all currently living organisms is the last universal ancestor, which lived about 3.9 billion years ago. The two earliest evidences for life on Earth are graphite found to be biogenic in 3.7 billion-year-old metasedimentary rocks discovered in western Greenland and microbial mat fossils found in 3.48 billion-year-old sandstone discovered in Western Australia. All currently living organisms on Earth share a common genetic heritage, though the suggestion of substantial horizontal gene transfer during early evolution has led to questions about the monophyly (single ancestry) of life. 6,331 groups of genes common to all living animals have been identified; these may have arisen from a single common ancestor that lived 650 million years ago in the Precambrian.

Universal common descent through an evolutionary process was first proposed by the British naturalist Charles Darwin in the concluding sentence of his 1859 book On the Origin of Species.

There is grandeur in this view of life, with its several powers, having been originally breathed into a few forms or into one; and that, whilst this planet has gone cycling on according to the fixed law of gravity, from so simple a beginning endless forms most beautiful and most wonderful have been, and are being, evolved.

In the 1740s, the French mathematician Pierre Louis Maupertuis made the first known suggestion that all organisms had a common ancestor, and had diverged through random variation and natural selection. In *Essai de cosmologie* (1750), Maupertuis noted:

May we not say that, in the fortuitous combination of the productions of Nature, since only those creatures *could* survive in whose organizations a certain degree of adaptation was present, there is nothing extraordinary in the fact that such adaptation is actually found in all these species which now exist? Chance, one might say, turned out a vast number of individuals; a small proportion of these were organized in such a manner that the animals' organs could satisfy their needs. A much greater number showed neither adaptation nor order; these last have all perished. Thus the species which we see today are but a small part of all those that a blind destiny has produced.

In 1790, the philosopher Immanuel Kant wrote in Kritik der Urteilskraft (Critique of Judgement) that the similarity of animal forms implies a common original type, and thus a common parent.

In 1794, Charles Darwin's grandfather, Erasmus Darwin asked:

Would it be too bold to imagine, that in the great length of time, since the earth began to exist, perhaps millions of ages before the commencement of the history of mankind, would it be too bold to imagine, that all warm-blooded animals have arisen from one living filament, which the great first cause endued with animality, with the power of acquiring new parts attended with new propensities, directed by irritations, sensations, volitions, and associations; and thus possessing the faculty of continuing to improve by its own inherent activity, and of delivering down those improvements by generation to its posterity, world without end?

Charles Darwin's views about common descent, as expressed in *On the Origin of Species*, were that it was probable that there was only one progenitor for all life forms:

Therefore I should infer from analogy that probably all the organic beings which have ever lived on this earth have descended from some one primordial form, into which life was first breathed.

But he precedes that remark by, "Analogy would lead me one step further, namely, to the belief that all animals and plants have descended from some one prototype. But analogy may be a deceitful guide." And in the subsequent edition, he asserts rather,

"We do not know all the possible transitional gradations between the simplest and the most perfect organs; it cannot be pretended that we know all the varied means of Distribution during the long lapse of years, or that we know how imperfect the Geological Record is. Grave as these several difficulties are, in my judgment they do not overthrow the theory of descent from a few created forms with subsequent modification".

Common descent was widely accepted amongst the scientific community after Darwin's publication. In 1907, Vernon Kellogg commented that "practically no naturalists of position and recognized attainment doubt the theory of descent."

In 2008, biologist T. Ryan Gregory noted that:

No reliable observation has ever been found to contradict the general notion of common descent. It should come as no surprise, then, that the scientific community at large has accepted evolutionary descent as a historical reality since Darwin's time and considers it among the most reliably established and fundamentally important facts in all of science.

Evidence

Common Biochemistry

All known forms of life are based on the same fundamental biochemical organization: genetic information encoded in DNA, transcribed into RNA, through the effect of protein- and RNA-enzymes, then translated into proteins by (highly similar) ribosomes, with ATP, NADPH and others as energy sources. Analysis of small sequence differences in widely shared substances such as cytochrome c further supports universal common descent. Some 23 proteins are found in all organisms, serving as enzymes carrying out core functions like DNA replication. The fact that only one such set of enzymes exists is convincing evidence of a single ancestry. 6,331 genes common to all living animals have been identified; these may have arisen from a single common ancestor that lived 650 million years ago in the Precambrian.

Common Genetic Code

Amino acids	nonpolar	polar	basic	acidic		Stop codon

Standard genetic code								
1st base	2nd base							
	T		C		A		G	
T	TTT	Phenyl-alanine	TCT	Serine	TAT	Tyrosine	TGT	Cysteine
	TTC		TCC		TAC		TGC	
	TTA	Leucine	TCA		TAA	Stop	TGA	Stop
	TTG		TCG		TAG	Stop	TGG	Tryptophan
C	CTT	Leucine	CCT	Proline	CAT	Histidine	CGT	Arginine
	CTC		CCC		CAC		CGC	
	CTA		CCA		CAA	Glutamine	CGA	
	CTG		CCG		CAG		CGG	
A	ATT	Isoleucine	ACT	Threonine	AAT	Asparagine	AGT	Serine
	ATC		ACC		AAC		AGC	
	ATA		ACA		AAA	Lysine	AGA	Arginine
	ATG	Methionine	ACG		AAG		AGG	
G	GTT	Valine	GCT	Alanine	GAT	Aspartic acid	GGT	Glycine
	GTC		GCC		GAC		GGC	
	GTA		GCA		GAA	Glutamic acid	GGA	
	GTG		GCG		GAG		GGG	

The genetic code is nearly identical for all known lifeforms, from bacteria and archaea to animals and plants. The universality of this code is generally regarded by biologists as definitive evidence in favor of universal common descent.

The way that codons (DNA triplets) are mapped to amino acids seems to be strongly optimised. Richard Egel argues that in particular the hydrophobic (non-polar) side-chains are well organised, suggesting that these enabled the earliest organisms to create peptides with water-repelling regions able to support the essential electron exchange (redox) reactions for energy transfer.

Selectively Neutral Similarities

Similarities which have no adaptive relevance cannot be explained by convergent evolution, and therefore they provide compelling support for universal common descent. Such evidence has come from two areas: amino acid sequences and DNA sequences. Proteins with the same three-dimensional structure need not have identical amino acid sequences; any irrelevant similarity between the sequences is evidence for common descent. In certain cases, there are several codons (DNA triplets) that code redundantly for the same amino acid. Since many species use the same codon at the same place to specify an amino acid that can be represented by more than one codon, that is evidence for their sharing a recent common ancestor. Had the amino acid sequences come from different ancestors, they would have been coded for by any of the redundant codons, and since the correct amino acids would already have been in place, natural selection would not have driven any change in the codons, however much time was available. Genetic drift could change the codons, but it would be extremely unlikely to make all the redundant codons in a whole sequence match exactly across multiple lineages. Similarly, shared nucleotide sequences, especially where these are apparently neutral such as the positioning of introns and pseudogenes, provide strong evidence of common ancestry.

Other Similarities

Biologists often point to the universality of many aspects of cellular life as supportive evidence to the more compelling evidence listed above. These similarities include the energy carrier adenosine triphosphate (ATP), and the fact that all amino acids found in proteins are left-handed. It is, however, possible that these similarities resulted because of the laws of physics and chemistry - rather than through universal common descent - and therefore resulted in convergent evolution. In contrast, there is evidence for homology of the central subunits of Transmembrane ATPases throughout all living organisms, especially how the rotating elements are bound to the membrane. This supports the assumption of a LUCA as a cellular organism, although primordial membranes may have been semipermeable and evolved later to the membranes of modern bacteria, and on a second path to those of modern archaea also.

Phylogenetic Trees

Another important piece of evidence is from detailed phylogenetic trees (i.e., "genealogic trees" of species) mapping out the proposed divisions and common ancestors of all living species. In 2010, Douglas L. Theobald published a statistical analysis of available genetic data, mapping them to phylogenetic trees, that gave "strong quantitative support, by a formal test, for the unity of life."

Traditionally, these trees have been built using morphological methods, such as appearance, embryology, etc. Recently, it has been possible to construct these trees using molecular data, based on similarities and differences between genetic and protein sequences. All these methods produce essentially similar results, even though most genetic variation has no influence over external morphology. That phylogenetic trees based on different types of information agree with each other is strong evidence of a real underlying common descent.

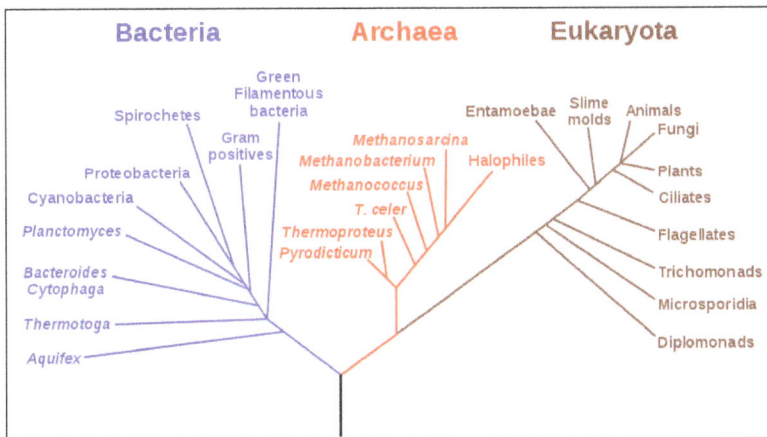

A phylogenetic tree based on ribosomal RNA genes implies a single origin for all life.

Potential Objections

Gene Exchange Clouds Phylogenetic Analysis

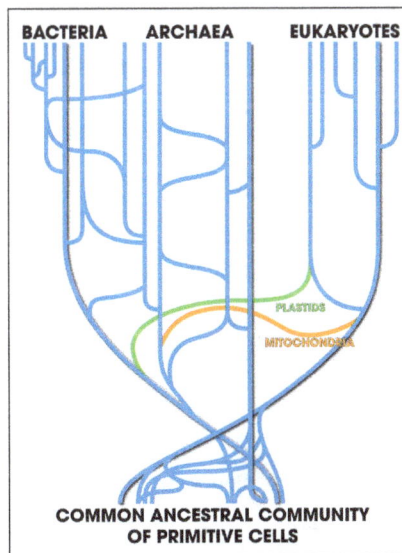

2005 tree of life shows many horizontal gene transfers, implying multiple possible origins.

Theobald noted that substantial horizontal gene transfer could have occurred during early evolution. Bacteria today remain capable of gene exchange between distantly-related lineages. This weakens the basic assumption of phylogenetic analysis, that similarity of genomes implies common ancestry, because sufficient gene exchange would allow lineages to share much of their genome whether or not they shared an ancestor (monophyly). This has led to questions about the single ancestry of life. However, biologists consider it very unlikely that completely unrelated proto-organisms could have exchanged genes, as their different coding mechanisms would have resulted only in garble rather than functioning systems. Later, however, many organisms all derived from a single ancestor could readily have shared genes that all worked in the same way, and it appears that they have.

Convergent Evolution

If early organisms had been driven by the same environmental conditions to evolve similar biochemistry convergently, they might independently have acquired similar genetic sequences. Theobald's "formal test" was accordingly criticised by Takahiro Yonezawa and colleagues for not including consideration of convergence. They argued that Theobald's test was insufficient to distinguish between the competing hypotheses. Theobald has defended his method against this claim, arguing that his tests distinguish between phylogenetic structure and mere sequence similarity. Therefore, Theobald argued, his results show that "real universally conserved proteins are homologous."

Speciation

Speciation is the evolutionary process by which populations evolve to become distinct species. The biologist Orator F. Cook coined the term in 1906 for cladogenesis, the splitting of lineages, as opposed to anagenesis, phyletic evolution within lineages. Charles Darwin was the first to describe the role of natural selection in speciation in his 1859 book *On the Origin of Species*. He also identified sexual selection as a likely mechanism, but found it problematic.

There are four geographic modes of speciation in nature, based on the extent to which speciating populations are isolated from one another: allopatric, peripatric, parapatric, and sympatric. Speciation may also be induced artificially, through animal husbandry, agriculture, or laboratory experiments. Whether genetic drift is a minor or major contributor to speciation is the subject matter of much ongoing discussion.

Rapid sympatric speciation can take place through polyploidy, such as by doubling of chromosome number; the result is progeny which are immediately reproductively isolated from the parent population. New species can also be created through hybridisation followed, if the hybrid is favoured by natural selection, by reproductive isolation.

Modes

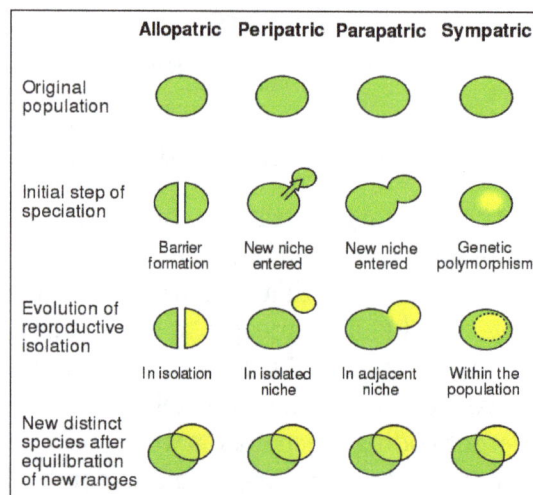

Comparison of allopatric, peripatric, parapatric and sympatric speciation.

All forms of natural speciation have taken place over the course of evolution; however, debate persists as to the relative importance of each mechanism in driving biodiversity.

One example of natural speciation is the diversity of the three-spined stickleback, a marine fish that, after the last glacial period, has undergone speciation into new freshwater colonies in isolated lakes and streams. Over an estimated 10,000 generations, the sticklebacks show structural differences that are greater than those seen between different genera of fish including variations in fins, changes in the number or size of their bony plates, variable jaw structure, and color differences.

Allopatric

During allopatric speciation, a population splits into two geographically isolated populations (for example, by habitat fragmentation due to geographical change such as mountain formation). The isolated populations then undergo genotypic or phenotypic divergence as: (a) they become subjected to dissimilar selective pressures; (b) they independently undergo genetic drift; (c) different mutations arise in the two populations. When the populations come back into contact, they have evolved such that they are reproductively isolated and are no longer capable of exchanging genes. Island genetics is the term associated with the tendency of small, isolated genetic pools to produce unusual traits. Examples include insular dwarfism and the radical changes among certain famous island chains, for example on Komodo. The Galápagos Islands are particularly famous for their influence on Charles Darwin. During his five weeks there he heard that Galápagos tortoises could be identified by island, and noticed that finches differed from one island to another, but it was only nine months later that he reflected that such facts could show that species were changeable. When he returned to England, his speculation on evolution deepened after experts informed him that these were separate species, not just varieties, and famously that other differing Galápagos birds were all species of finches. Though the finches were less important for Darwin, more recent research has shown the birds now known as Darwin's finches to be a classic case of adaptive evolutionary radiation.

Peripatric

In peripatric speciation, a subform of allopatric speciation, new species are formed in isolated, smaller peripheral populations that are prevented from exchanging genes with the main population. It is related to the concept of a founder effect, since small populations often undergo bottlenecks. Genetic drift is often proposed to play a significant role in peripatric speciation.

Case studies include Mayr's investigation of bird fauna; the Australian bird Petroica multicolor; and reproductive isolation in populations of Drosophila subject to population bottlenecking.

Parapatric

In parapatric speciation, there is only partial separation of the zones of two diverging populations afforded by geography; individuals of each species may come in contact or cross habitats from time to time, but reduced fitness of the heterozygote leads to selection for behaviours or mechanisms that prevent their interbreeding. Parapatric speciation is modelled on continuous variation within a "single," connected habitat acting as a source of natural selection rather than the effects of isolation of habitats produced in peripatric and allopatric speciation.

Parapatric speciation may be associated with differential landscape-dependent selection. Even if there is a gene flow between two populations, strong differential selection may impede assimilation and different species may eventually develop. Habitat differences may be more important in the development of reproductive isolation than the isolation time. Caucasian rock lizards Darevskia rudis, D. valentini and D. portschinskii all hybridize with each other in their hybrid zone; however, hybridization is stronger between D. portschinskii and D. rudis, which separated earlier but live in similar habitats than between D. valentini and two other species, which separated later but live in climatically different habitats.

Ecologists refer to parapatric and peripatric speciation in terms of ecological niches. A niche must be available in order for a new species to be successful. Ring species such as Larus gulls have been claimed to illustrate speciation in progress, though the situation may be more complex. The grass Anthoxanthum odoratum may be starting parapatric speciation in areas of mine contamination.

Sympatric

Cichlids such as Haplochromis nyererei diversified
by sympatric speciation in the Rift Valley lakes.

Sympatric speciation is the formation of two or more descendant species from a single ancestral species all occupying the same geographic location.

Often-cited examples of sympatric speciation are found in insects that become dependent on different host plants in the same area.

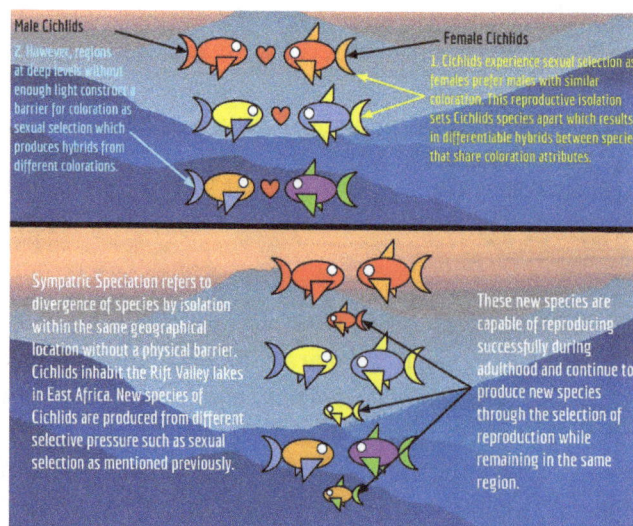

Sympatric Speciation with Cichlids.

The best known example of sympatric speciation is that of the cichlids of East Africa inhabiting the Rift Valley lakes, particularly Lake Victoria, Lake Malawi and Lake Tanganyika. There are over 800 described species, and according to estimates, there could be well over 1,600 species in the region. Their evolution is cited as an example of both natural and sexual selection. A 2008 study suggests that sympatric speciation has occurred in Tennessee cave salamanders. Sympatric speciation driven by ecological factors may also account for the extraordinary diversity of crustaceans living in the depths of Siberia's Lake Baikal.

Budding speciation has been proposed as a particular form of sympatric speciation, whereby small groups of individuals become progressively more isolated from the ancestral stock by breeding preferentially with one another. This type of speciation would be driven by the conjunction of various advantages of inbreeding such as the expression of advantageous recessive phenotypes, reducing the recombination load, and reducing the cost of sex.

Rhagoletis pomonella, the hawthorn fly, appears
to be in the process of sympatric speciation.

The hawthorn fly (Rhagoletis pomonella), also known as the apple maggot fly, appears to be undergoing sympatric speciation. Different populations of hawthorn fly feed on different fruits. A distinct population emerged in North America in the 19th century some time after apples, a non-native species, were introduced. This apple-feeding population normally feeds only on apples and not on the historically preferred fruit of hawthorns. The current hawthorn feeding population does not normally feed on apples. Some evidence, such as that six out of thirteen allozyme loci are different, that hawthorn flies mature later in the season and take longer to mature than apple flies; and that there is little evidence of interbreeding (researchers have documented a 4-6% hybridization rate) suggests that sympatric speciation is occurring.

Methods of Selection

Reinforcement

Reinforcement, sometimes referred to as the Wallace effect, is the process by which natural selection increases reproductive isolation. It may occur after two populations of the same species are separated and then come back into contact. If their reproductive isolation was complete, then they will have already developed into two separate incompatible species. If their reproductive isolation is incomplete, then further mating between the populations will produce hybrids, which may or may not be fertile. If the hybrids are infertile, or fertile but less fit than their ancestors, then there will be further reproductive isolation and speciation has essentially occurred (e.g., as in horses and donkeys).

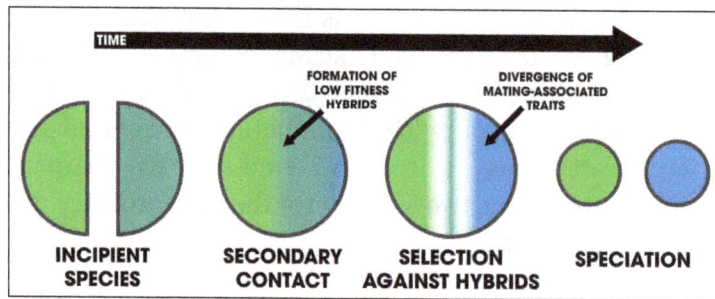

Reinforcement assists speciation by selecting against hybrids.

The reasoning behind this is that if the parents of the hybrid offspring each have naturally selected traits for their own certain environments, the hybrid offspring will bear traits from both, therefore would not fit either ecological niche as well as either parent. The low fitness of the hybrids would cause selection to favor assortative mating, which would control hybridization. This is sometimes called the Wallace effect after the evolutionary biologist Alfred Russel Wallace who suggested in the late 19th century that it might be an important factor in speciation. Conversely, if the hybrid offspring are more fit than their ancestors, then the populations will merge back into the same species within the area they are in contact.

Reinforcement favoring reproductive isolation is required for both parapatric and sympatric speciation. Without reinforcement, the geographic area of contact between different forms of the same species, called their "hybrid zone," will not develop into a boundary between the different species. Hybrid zones are regions where diverged populations meet and interbreed. Hybrid offspring are very common in these regions, which are usually created by diverged species coming into secondary contact. Without reinforcement, the two species would have uncontrollable inbreeding. Reinforcement may be induced in artificial selection experiments.

Ecological

Ecological selection is "the interaction of individuals with their environment during resource acquisition". Natural selection is inherently involved in the process of speciation, whereby, "under ecological speciation, populations in different environments, or populations exploiting different resources, experience contrasting natural selection pressures on the traits that directly or indirectly bring about the evolution of reproductive isolation". Evidence for the role ecology plays in the process of speciation exists. Studies of stickleback populations support ecologically-linked speciation arising as a by-product, alongside numerous studies of parallel speciation, where isolation evolves between independent populations of species adapting to contrasting environments than between independent populations adapting to similar environments. Ecological speciation occurs with much of the evidence, "accumulated from top-down studies of adaptation and reproductive isolation".

Sexual Selection

It is widely appreciated that sexual selection could drive speciation in many clades, independently of natural selection. However the term "speciation", in this context, tends to be used in two different, but not mutually exclusive senses. The first and most commonly used sense refers to the "birth" of new species. That is, the splitting of an existing species into two separate species, or the budding off of a new species from a parent species, both driven by a biological "fashion fad" (a preference for a feature, or features, in one or both sexes, that do not necessarily have any adaptive qualities). In the second sense, "speciation" refers to the wide-spread tendency of sexual creatures to be grouped into clearly defined species, rather than forming a continuum of phenotypes both in time and space - which would be the more obvious or logical consequence of natural selection. This was indeed recognized by Darwin as problematic, and included in his *On the Origin of Species* (1859), under the heading "Difficulties with the Theory". There are several suggestions as to how mate choice might play a significant role in resolving Darwin's dilemma.

Artificial Speciation

Gaur (Indian bison) can interbreed with domestic cattle.

New species have been created by animal husbandry, but the dates and methods of the initiation of such species are not clear. Often, the domestic counterpart of the wild ancestor can still interbreed and produce fertile offspring as in the case of domestic cattle, that can be considered the same species as several varieties of wild ox, gaur, yak, etc., or domestic sheep that can interbreed with the mouflon.

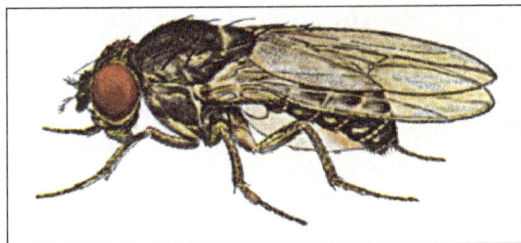

Male Drosophila pseudoobscura.

The best-documented creations of new species in the laboratory were performed in the late 1980s. William R. Rice and George W. Salt bred Drosophila melanogaster fruit flies using a maze with three different choices of habitat such as light/dark and wet/dry. Each generation was placed into the maze, and the groups of flies that came out of two of the eight exits were set apart to breed with each other in their respective groups. After thirty-five generations, the two groups and their offspring were isolated reproductively because of their strong habitat preferences: they mated only within the areas they preferred, and so did not mate with flies that preferred the other areas. The history of such attempts is described by Rice and Elen E. Hostert. Diane Dodd used a laboratory experiment to show how reproductive isolation can develop in Drosophila pseudoobscura fruit flies after several generations by placing them in different media, starch- and maltose-based media.

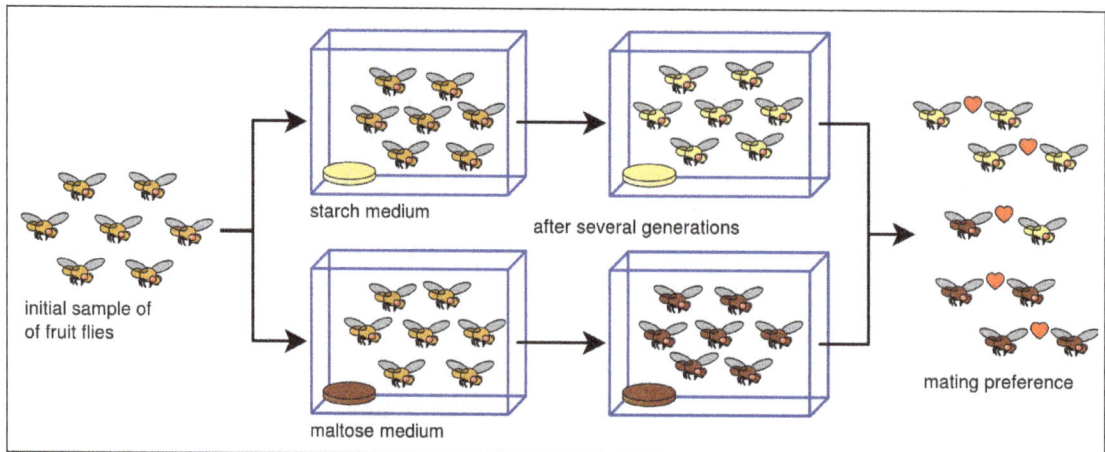

Dodd's experiment has been easy for many others to replicate, including with other kinds of fruit flies and foods. Research in 2005 has shown that this rapid evolution of reproductive isolation may in fact be a relic of infection by Wolbachia bacteria.

Alternatively, these observations are consistent with the notion that sexual creatures are inherently reluctant to mate with individuals whose appearance or behavior is different from the norm. The risk that such deviations are due to heritable maladaptations is very high. Thus, if a sexual creature, unable to predict natural selection's future direction, is conditioned to produce the fittest offspring possible, it will avoid mates with unusual habits or features. Sexual creatures will then inevitably tend to group themselves into reproductively isolated species.

Genetics

Few speciation genes have been found. They usually involve the reinforcement process of late stages of speciation. In 2008, a speciation gene causing reproductive isolation was reported. It causes hybrid sterility between related subspecies. The order of speciation of three groups from a common ancestor may be unclear or unknown; a collection of three such species is referred to as a "trichotomy."

Speciation via Polyploidy

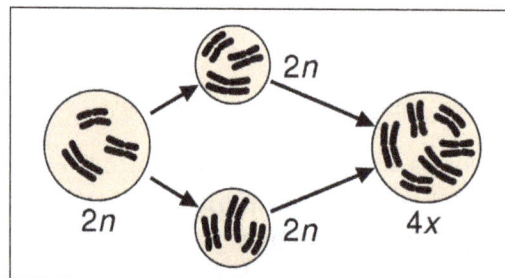

Speciation via polyploidy: A diploid cell undergoes failed meiosis, producing diploid gametes, which self-fertilize to produce a tetraploid zygote. In plants, this can effectively be a new species, reproductively isolated from its parents, and able to reproduce.

Polyploidy is a mechanism that has caused many rapid speciation events in sympatry because offspring of, for example, tetraploid x diploid matings often result in triploid sterile progeny. However, not all polyploids are reproductively isolated from their parental plants, and gene flow may still occur for example through triploid hybrid x diploid matings that produce tetraploids, or matings between meiotically unreduced gametes from diploids and gametes from tetraploids.

It has been suggested that many of the existing plant and most animal species have undergone an event of polyploidization in their evolutionary history. Reproduction of successful polyploid species is sometimes asexual, by parthenogenesis or apomixis, as for unknown reasons many asexual organisms are polyploid. Rare instances of polyploid mammals are known, but most often result in prenatal death.

Hybrid Speciation

Hybridization between two different species sometimes leads to a distinct phenotype. This phenotype can also be fitter than the parental lineage and as such natural selection may then favor these individuals. Eventually, if reproductive isolation is achieved, it may lead to a separate species. However, reproductive isolation between hybrids and their parents is particularly difficult to achieve and thus hybrid speciation is considered an extremely rare event. The Mariana mallard is thought to have arisen from hybrid speciation.

Hybridization is an important means of speciation in plants, since polyploidy (having more than two copies of each chromosome) is tolerated in plants more readily than in animals. Polyploidy is important in hybrids as it allows reproduction, with the two different sets of chromosomes each being able to pair with an identical partner during meiosis. Polyploids also have more genetic diversity, which allows them to avoid inbreeding depression in small populations.

Hybridization without change in chromosome number is called homoploid hybrid speciation. It is considered very rare but has been shown in *Heliconius* butterflies and sunflowers. Polyploid speciation, which involves changes in chromosome number, is a more common phenomenon, especially in plant species.

Gene Transposition

Theodosius Dobzhansky, who studied fruit flies in the early days of genetic research in 1930s, speculated that parts of chromosomes that switch from one location to another might cause a species to split into two different species. He mapped out how it might be possible for sections of chromosomes to relocate themselves in a genome. Those mobile sections can cause sterility in inter-species hybrids, which can act as a speciation pressure. In theory, his idea was sound, but scientists long debated whether it actually happened in nature. Eventually a competing theory involving the gradual accumulation of mutations was shown to occur in nature so often that geneticists largely dismissed the moving gene hypothesis. However, 2006 research shows that jumping of a gene from one chromosome to another can contribute to the birth of new species. This validates the reproductive isolation mechanism, a key component of speciation.

Rates

There is debate as to the rate at which speciation events occur over geologic time. While some evolutionary biologists claim that speciation events have remained relatively constant and gradual over time (known as "Phyletic gradualism"), some palaeontologists such as Niles Eldredge and Stephen Jay Gould have argued that species usually remain unchanged over long stretches of time, and that speciation occurs only over relatively brief intervals, a view known as punctuated equilibrium.

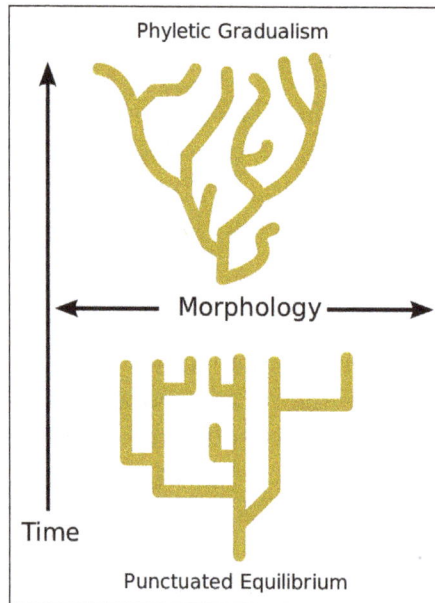

Phyletic gradualism, above, consists of relatively slow change over geological time. Punctuated equilibrium, bottom, consists of morphological stability and rare, relatively rapid bursts of evolutionary change.

Punctuated Evolution

Evolution can be extremely rapid, as shown in the creation of domesticated animals and plants in a very short geological space of time, spanning only a few tens of thousands of years. Maize (Zea mays), for instance, was created in Mexico in only a few thousand years, starting about 7,000 to 12,000 years ago. This raises the question of why the long term rate of evolution is far slower than is theoretically possible.

Plants and domestic animals can differ markedly from their wild ancestors:

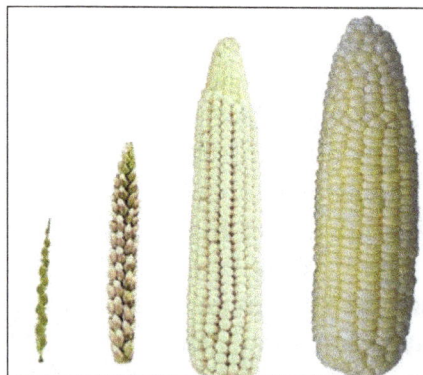

Evolution is imposed on species or groups. It is not planned or striven for in some Lamarckist way. The mutations on which the process depends are random events, and, except for the "silent mutations" which do not affect the functionality or appearance of the carrier, are thus usually disadvantageous, and their chance of proving to be useful in the future is vanishingly small. Therefore, while a species or group might benefit from being able to adapt to a new environment by accumulating a wide range of genetic variation, this is to the detriment of the individuals who have to carry these mutations until a small, unpredictable minority of them ultimately contributes to such an adaptation. Thus, the capability to evolve would require group selection, a concept discredited by (for example) George C. Williams, John Maynard Smith and Richard Dawkins as selectively disadvantageous to the individual.

Ancestral wild cabbage.

Domesticated cauliflower.

The resolution to Darwin's second dilemma might thus come about as follows:

If sexual individuals are disadvantaged by passing mutations on to their offspring, they will avoid mutant mates with strange or unusual characteristics. Mutations that affect the external appearance of their carriers will then rarely be passed on to the next and subsequent generations. They would therefore seldom be tested by natural selection. Evolution is, therefore, effectively halted or slowed down considerably. The only mutations that can accumulate in a population, on this punctuated equilibrium view, are ones that have no noticeable effect on the outward appearance and functionality of their bearers (i.e., they are "silent" or "neutral mutations," which can be, and are, used to trace the relatedness and age of populations and species). This argument implies that evolution can only occur if mutant mates cannot be avoided, as a result of a severe scarcity of potential mates. This is most likely to occur in small, isolated communities. These occur most commonly on small islands, in remote valleys, lakes, river systems, or caves, or during the aftermath of a mass extinction. Under these circumstances, not only is the choice of mates severely restricted but population bottlenecks, founder effects, genetic drift and inbreeding cause rapid, random changes in the isolated population's genetic composition. Furthermore, hybridization with a related species trapped in the same isolate might introduce additional genetic changes. If an isolated population such as this survives its genetic upheavals, and subsequently expands into an unoccupied niche, or into a niche in which it has an advantage over its competitors, a new species, or subspecies, will have come in being. In geological terms, this will be an abrupt event. A resumption of avoiding mutant mates will thereafter result, once again, in evolutionary stagnation.

Ancestral Prussian carp.

Domestic goldfish.

Ancestral mouflon.

Domestic sheep.

In apparent confirmation of this punctuated equilibrium view of evolution, the fossil record of an evolutionary progression typically consists of species that suddenly appear, and ultimately disappear, hundreds of thousands or millions of years later, without any change in external appearance. Graphically, these fossil species are represented by lines parallel with the time axis, whose lengths depict how long each of them existed. The fact that the lines remain parallel with the time axis illustrates the unchanging appearance of each of the fossil species depicted on the graph. During each species' existence new species appear at random intervals, each also lasting many hundreds of thousands of years before disappearing without a change in appearance. The exact relatedness of these concurrent species is generally impossible to determine. This is illustrated in the diagram depicting the distribution of hominin species through time since the hominins separated from the line that led to the evolution of our closest living primate relatives, the chimpanzees.

For similar evolutionary time lines see, for instance, the paleontological list of African dinosaurs, Asian dinosaurs, the Lampriformes and Amiiformes.

Adaptation

Adaptation has three related meanings. Firstly, it is the dynamic evolutionary process that fits organisms to their environment, enhancing their evolutionary fitness. Secondly, it is a state reached by the population during that process. Thirdly, it is a phenotypic trait or adaptive trait, with a functional role in each individual organism, that is maintained and has evolved through natural selection.

Historically, adaptation has been described from the time of the ancient philosophers such as Empedocles and Aristotle. In 18th and 19th century natural theology, adaptation was taken as evidence for the existence of a deity. Charles Darwin proposed instead that it was explained by natural selection.

Adaptation is related to biological fitness, which governs the rate of evolution, as measured by change in gene frequencies. Often, two or more species co-adapt and co-evolve as they develop adaptations that interlock with those of the other species, such as with flowering plants and pollinating insects. In mimicry, species evolve to resemble other species; in Müllerian mimicry this is a mutually beneficial co-evolution as each of a group of strongly defended species (such as wasps able to sting) come to advertise their defences in the same way. Features evolved for one purpose may be co-opted for a different one, as when the insulating feathers of dinosaurs were co-opted for bird flight.

Adaptation is a major topic in the philosophy of biology, as it concerns function and purpose (teleology). Some biologists try to avoid terms which imply purpose in adaptation, not least because it suggests a deity's intentions, but others note that adaptation is necessarily purposeful.

Principles

The significance of an adaptation can only be understood in relation to the total biology of the species.

—Julian Huxley, Evolution: The Modern Synthesis

Adaptation is primarily a process rather than a physical form or part of a body. An internal parasite (such as a liver fluke) can illustrate the distinction: such a parasite may have a very simple bodily structure, but nevertheless the organism is highly adapted to its specific environment. From this we see that adaptation is not just a matter of visible traits: in such parasites critical adaptations take place in the life cycle, which is often quite complex. However, as a practical term, "adaptation" often refers to a product: those features of a species which result from the process. Many aspects of an animal or plant can be correctly called adaptations, though there are always some features whose function remains in doubt. By using the term adaptation for the evolutionary process, and adaptive trait for the bodily part or function (the product), one may distinguish the two different senses of the word.

Adaptation is one of the two main processes that explain the observed diversity of species, such as the different species of Darwin's finches. The other process is speciation, in which new species arise, typically through reproductive isolation. A favourite example used today to study the interplay of adaptation and speciation is the evolution of cichlid fish in African lakes, where the question of reproductive isolation is complex.

Adaptation is not always a simple matter where the ideal phenotype evolves for a given external environment. An organism must be viable at all stages of its development and at all stages of its evolution. This places constraints on the evolution of development, behaviour, and structure of organisms. The main constraint, over which there has been much debate, is the requirement that each genetic and phenotypic change during evolution should be relatively small, because developmental systems are so complex and interlinked. However, it is not clear what "relatively small" should mean, for example polyploidy in plants is a reasonably common large genetic change. The origin of eukaryotic endosymbiosis is a more dramatic example.

All adaptations help organisms survive in their ecological niches. The adaptive traits may be structural, behavioural or physiological. Structural adaptations are physical features of an organism, such as shape, body covering, armament, and internal organization. Behavioural adaptations are inherited systems of behaviour, whether inherited in detail as instincts, or as a neuropsychological capacity for learning. Examples include searching for food, mating, and vocalizations. Physiological adaptations permit the organism to perform special functions such as making venom, secreting slime, and phototropism), but also involve more general functions such as growth and development, temperature regulation, ionic balance and other aspects of homeostasis. Adaptation affects all aspects of the life of an organism.

The following definitions are given by the evolutionary biologist Theodosius Dobzhansky:

- Adaptation is the evolutionary process whereby an organism becomes better able to live in its habitat or habitats.

- Adaptedness is the state of being adapted: the degree to which an organism is able to live and reproduce in a given set of habitats.

- An adaptive trait is an aspect of the developmental pattern of the organism which enables or enhances the probability of that organism surviving and reproducing.

Some generalists, such as birds, have the
flexibility to adapt to urban areas.

Adaptation differs from flexibility, acclimatization, and learning, all of which are changes during life which are not inherited. Flexibility deals with the relative capacity of an organism to maintain itself in different habitats: its degree of specialization. Acclimatization describes automatic physiological adjustments during life; learning means improvement in behavioral performance during life.

Flexibility stems from phenotypic plasticity, the ability of an organism with a given genotype to change its phenotype in response to changes in its habitat, or to move to a different habitat. The degree of flexibility is inherited, and varies between individuals. A highly specialized animal or plant lives only in a well-defined habitat, eats a specific type of food, and cannot survive if its needs are not met. Many herbivores are like this; extreme examples are koalas which depend on *Eucalyptus*, and giant pandas which require bamboo. A generalist, on the other hand, eats a range of food, and can survive in many different conditions. Examples are humans, rats, crabs and many carnivores. The *tendency* to behave in a specialized or exploratory manner is inherited—it is an adaptation. Rather different is developmental flexibility: "An animal or plant is developmentally flexible if when it is raised in or transferred to new conditions, it changes in structure so that it is better fitted to survive in the new environment," writes evolutionary biologist John Maynard Smith.

If humans move to a higher altitude, respiration and physical exertion become a problem, but after spending time in high altitude conditions they acclimatize to the reduced partial pressure of oxygen, such as by producing more red blood cells. The ability to acclimatize is an adaptation, but the acclimatization itself is not. Fecundity goes down, but deaths from some tropical diseases also go down. Over a longer period of time, some people are better able to reproduce at high altitudes than others. They contribute more heavily to later generations, and gradually by natural selection the whole population becomes adapted to the new conditions. This has demonstrably occurred, as the observed performance of long-term communities at higher altitude is significantly better than the performance of new arrivals, even when the new arrivals have had time to acclimatize.

Adaptedness and Fitness

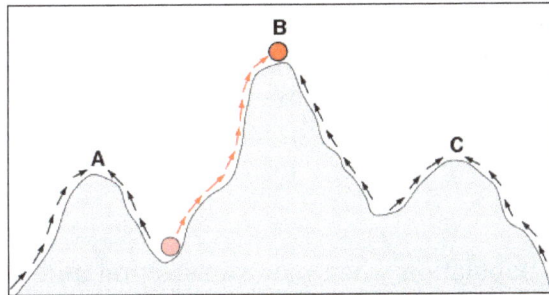

In this sketch of a fitness landscape, a population can evolve by following
the arrows to the adaptive peak at point B, and the points A and C are local
optima where a population could become trapped.

There is a relationship between adaptedness and the concept of fitness used in population genetics. Differences in fitness between genotypes predict the rate of evolution by natural selection. Natural selection changes the relative frequencies of alternative phenotypes, insofar as they are heritable. However, a phenotype with high adaptedness may not have high fitness. Dobzhansky mentioned the example of the Californian redwood, which is highly adapted, but a relict species in danger of extinction. Elliott Sober commented that adaptation was a retrospective concept since it implied something about the history of a trait, whereas fitness predicts a trait's future.

- Relative fitness: The average contribution to the next generation by a genotype or a class of genotypes, relative to the contributions of other genotypes in the population. This is also known as Darwinian fitness, selection coefficient, and other terms.

- Absolute fitness: The absolute contribution to the next generation by a genotype or a class of genotypes. Also known as the Malthusian parameter when applied to the population as a whole.

- Adaptedness: The extent to which a phenotype fits its local ecological niche. Researchers can sometimes test this through a reciprocal transplant.

Sewall Wright proposed that populations occupy *adaptive peaks* on a fitness landscape. To evolve to another, higher peak, a population would first have to pass through a valley of maladaptive intermediate stages, and might be "trapped" on a peak that is not optimally adapted.

Types

Adaptation is the heart and soul of evolution.

—Niles Eldredge, Reinventing Darwin: The Great Debate at the High Table of Evolutionary Theory.

Changes in Habitat

Before Darwin, adaptation was seen as a fixed relationship between an organism and its habitat. It was not appreciated that as the climate changed, so did the habitat; and as the habitat changed, so did the biota. Also, habitats are subject to changes in their biota: for example, invasions of species from other areas. The relative numbers of species in a given habitat are always changing. Change is the rule, though much depends on the speed and degree of the change. When the habitat changes, three main things may happen to a resident population: habitat tracking, genetic change or extinction. In fact, all three things may occur in sequence. Of these three effects only genetic change brings about adaptation. When a habitat changes, the resident population typically moves to more suitable places; this is the typical response of flying insects or oceanic organisms, which have wide (though not unlimited) opportunity for movement. This common response is called *habitat tracking*. It is one explanation put forward for the periods of apparent stasis in the fossil record (the punctuated equilibrium theory).

Genetic Change

Genetic change occurs in a population when natural selection and mutations act on its genetic variability. The first pathways of enzyme-based metabolism may have been parts of purine nucleotide metabolism, with previous metabolic pathways being part of the ancient RNA world. By this means, the population adapts genetically to its circumstances. Genetic changes may result in visible structures, or may adjust physiological activity in a way that suits the habitat.

Habitats and biota do frequently change. Therefore, it follows that the process of adaptation is never finally complete. Over time, it may happen that the environment changes little, and the species comes to fit its surroundings better and better. On the other hand, it may happen that changes in the environment occur relatively rapidly, and then the species becomes less and less well adapted. Seen like this, adaptation is a genetic tracking process, which goes on all the time to some extent, but especially when the population cannot or does not move to another, less hostile area. Given enough genetic change, as well as specific demographic conditions, an adaptation may be enough to bring a population back from the brink of extinction in a process called evolutionary rescue. Adaptation does affect, to some extent, every species in a particular ecosystem.

Leigh Van Valen thought that even in a stable environment, competing species constantly had to adapt to maintain their relative standing. This became known as the Red Queen hypothesis, as seen in host-parasite interaction.

Existing genetic variation and mutation were the traditional sources of material on which natural selection could act. In addition, horizontal gene transfer is possible between organisms in different species, using mechanisms as varied as gene cassettes, plasmids, transposons and viruses such as bacteriophages.

Co-adaptation

In coevolution, where the existence of one species is tightly bound up with the life of another species, new or 'improved' adaptations which occur in one species are often followed by the appearance and spread of corresponding features in the other species. These co-adaptational relationships are intrinsically dynamic, and may continue on a trajectory for millions of years, as has occurred in the relationship between flowering plants and pollinating insects.

Pollinating insects are co-adapted with flowering plants.

Mimicry

Bates' work on Amazonian butterflies led him to develop the first scientific account of mimicry, especially the kind of mimicry which bears his name: Batesian mimicry. This is the mimicry by a palatable species of an unpalatable or noxious species (the model), gaining a selective advantage as predators avoid the model and therefore also the mimic. Mimicry is thus an anti-predator adaptation. A common example seen in temperate gardens is the hoverfly, many of which—though bearing no sting—mimic the warning coloration of hymenoptera (wasps and bees). Such mimicry does not need to be perfect to improve the survival of the palatable species.

A and B show real wasps; the rest are Batesian
mimics: three hoverflies and one beetle.

Bates, Wallace and Fritz Müller believed that Batesian and Müllerian mimicry provided evidence for the action of natural selection, a view which is now standard amongst biologists.

Trade-offs

It is a profound truth that Nature does not know best; that genetical evolution is a story of waste, makeshift, compromise and blunder.

— Peter Medawar, The Future of Man.

All adaptations have a downside: horse legs are great for running on grass, but they can't scratch their backs; mammals' hair helps temperature, but offers a niche for ectoparasites; the only flying penguins do is under water. Adaptations serving different functions may be mutually destructive. Compromise and makeshift occur widely, not perfection. Selection pressures pull in different directions, and the adaptation that results is some kind of compromise.

Since the phenotype as a whole is the target of selection, it is impossible to improve simultaneously all aspects of the phenotype to the same degree.

— Ernst Mayr, The Growth of Biological Thought: Diversity, Evolution, and Inheritance.

Consider the antlers of the Irish elk, (often supposed to be far too large; in deer antler size has an allometric relationship to body size). Obviously, antlers serve positively for defence against predators, and to score victories in the annual rut. But they are costly in terms of resource. Their size during the last glacial period presumably depended on the relative gain and loss of reproductive capacity in the population of elks during that time. As another example, camouflage to avoid detection is destroyed when vivid coloration is displayed at mating time. Here the risk to life is counterbalanced by the necessity for reproduction.

Stream-dwelling salamanders, such as Caucasian salamander or Gold-striped salamander have very slender, long bodies, perfectly adapted to life at the banks of fast small rivers and mountain brooks. Elongated body protects their larvae from being washed out by current. However, elongated body increases risk of desiccation and decreases dispersal ability of the salamanders; it also negatively affects their fecundity. As a result, fire salamander, less perfectly adapted to the mountain brook habitats, is in general more successful, have a higher fecundity and broader geographic range.

An Indian peacock's train in full display.

The peacock's ornamental train (grown anew in time for each mating season) is a famous adaptation. It must reduce his maneuverability and flight, and is hugely conspicuous; also, its growth costs food resources. Darwin's explanation of its advantage was in terms of sexual selection: "This depends on the advantage which certain individuals have over other individuals of the same sex and species, in exclusive relation to reproduction." The kind of sexual selection represented by the peacock is called 'mate choice,' with an implication that the process selects the more fit over the less fit, and so has survival value. The recognition of sexual selection was for a long time in abeyance, but has been rehabilitated.

The conflict between the size of the human foetal brain at birth, (which cannot be larger than about 400 cm³, else it will not get through the mother's pelvis) and the size needed for an adult brain (about 1400 cm³), means the brain of a newborn child is quite immature. The most vital things in human life (locomotion, speech) just have to wait while the brain grows and matures. That is the result of the birth compromise. Much of the problem comes from our upright bipedal stance, without which our pelvis could be shaped more suitably for birth. Neanderthals had a similar problem.

As another example, the long neck of a giraffe brings benefits but at a cost. The neck of a giraffe can be up to 2 m (6 ft 7 in) in length. The benefits are that it can be used for inter-species competition or for foraging on tall trees where shorter herbivores cannot reach. The cost is that a long neck is heavy and adds to the animal's body mass, requiring additional energy to build the neck and to carry its weight around.

Shifts in Function

Adaptation and function are two aspects of one problem.

—Julian Huxley, Evolution: The Modern Synthesis

Pre-adaptation

Pre-adaptation occurs when a population has characteristics which by chance are suited for a set of conditions not previously experienced. For example, the polyploid cordgrass *Spartina townsendii* is better adapted than either of its parent species to their own habitat of saline marsh and mudflats. Among domestic animals, the White Leghorn chicken is markedly more resistant to vitamin B_1 deficiency than other breeds; on a plentiful diet this makes no difference, but on a restricted diet this preadaptation could be decisive.

Pre-adaptation may arise because a natural population carries a huge quantity of genetic variability. In diploid eukaryotes, this is a consequence of the system of sexual reproduction, where mutant alleles get partially shielded, for example, by genetic dominance. Microorganisms, with their huge populations, also carry a great deal of genetic variability. The first experimental evidence of the pre-adaptive nature of genetic variants in microorganisms was provided by Salvador Luria and Max Delbrück who developed the Fluctuation Test, a method to show the random fluctuation of pre-existing genetic changes that conferred resistance to bacteriophages in Escherichia coli.

Co-option of Existing Traits: Exaptation

Features that now appear as adaptations sometimes arose by co-option of existing traits, evolved for some other purpose. The classic example is the ear ossicles of mammals, which we know from paleontological and embryological evidence originated in the upper and lower jaws and the hyoid bone of their synapsid ancestors, and further back still were part of the gill arches of early fish. The word exaptation was coined to cover these common evolutionary shifts in function. The flight feathers of birds evolved from the much earlier feathers of dinosaurs, which might have been used for insulation or for display.

The feathers of Sinosauropteryx, a dinosaur with feathers,
were used for insulation, making them an exaptation for flight.

Non-adaptive Traits

Some traits do not appear to be adaptive, that is, they have a neutral or deleterious effect on fitness in the current environment. Because genes often have pleiotropic effects, not all traits may be functional: they may be what Stephen Jay Gould and Richard Lewontin called spandrels, features brought about by neighbouring adaptations, on the analogy with the often highly decorated triangular areas between pairs of arches in architecture, which began as functionless features.

Another possibility is that a trait may have been adaptive at some point in an organism's evolutionary history, but a change in habitats caused what used to be an adaptation to become unnecessary or even maladapted. Such adaptations are termed vestigial. Many organisms have vestigial organs, which are the remnants of fully functional structures in their ancestors. As a result of changes in lifestyle the organs became redundant, and are either not functional or reduced in functionality. Since any structure represents some kind of cost to the general economy of the body, an advantage may accrue from their elimination once they are not functional. Examples: wisdom teeth in humans; the loss of pigment and functional eyes in cave fauna; the loss of structure in endoparasites.

Extinction and Coextinction

If a population cannot move or change sufficiently to preserve its long-term viability, then obviously, it will become extinct, at least in that locale. The species may or may not survive in other locales. Species extinction occurs when the death rate over the entire species exceeds the birth rate for a long enough period for the species to disappear. It was an observation of Van Valen that groups of species tend to have a characteristic and fairly regular rate of extinction.

Just as there is co-adaptation, there is also coextinction, the loss of a species due to the extinction of another with which it is coadapted, as with the extinction of a parasitic insect following the loss of its host, or when a flowering plant loses its pollinator, or when a food chain is disrupted.

Heredity

Heredity, also called inheritance or biological inheritance, is the passing on of traits from parents to their offspring; either through asexual reproduction or sexual reproduction, the offspring cells or organisms acquire the genetic information of their parents. Through heredity, variations between individuals can accumulate and cause species to evolve by natural selection. The study of heredity in biology is genetics.

Heredity of phenotypic traits: Father and son
with prominent ears and crowns.

In humans, eye color is an example of an inherited characteristic: an individual might inherit the "brown-eye trait" from one of the parents. Inherited traits are controlled by genes and the complete set of genes within an organism's genome is called its genotype.

The complete set of observable traits of the structure and behavior of an organism is called its phenotype. These traits arise from the interaction of its genotype with the environment. As a result, many aspects of an organism's phenotype are not inherited. For example, suntanned skin comes from the interaction between a person's phenotype and sunlight; thus, suntans are not passed on to people's children. However, some people tan more easily than others, due to differences in their genotype: a striking example is people with the inherited trait of albinism, who do not tan at all and are very sensitive to sunburn.

Heritable traits are known to be passed from one generation to the next via DNA, a molecule that encodes genetic information. DNA is a long polymer that incorporates four types of bases, which are interchangeable. The sequence of bases along a particular DNA molecule specifies the genetic information: this is comparable to a sequence of letters spelling out a passage of text. Before a cell divides through mitosis, the DNA is copied, so that each of the resulting two cells will inherit the DNA sequence. A portion of a DNA molecule that specifies a single functional unit is called a gene; different genes have different sequences of bases. Within cells, the long strands of DNA form condensed structures called chromosomes. Organisms inherit genetic material from their parents in the form of homologous chromosomes, containing a unique combination of DNA sequences that code for genes. The specific location of a DNA sequence within a chromosome is known as a locus. If the DNA sequence at a particular locus varies between individuals, the different forms of this sequence are called alleles. DNA sequences can change through mutations, producing new alleles. If a mutation occurs within a gene, the new allele may affect the trait that the gene controls, altering the phenotype of the organism.

However, while this simple correspondence between an allele and a trait works in some cases, most traits are more complex and are controlled by multiple interacting genes within and among organisms. Developmental biologists suggest that complex interactions in genetic networks and communication among cells can lead to heritable variations that may underlie some of the mechanics in developmental plasticity and canalization.

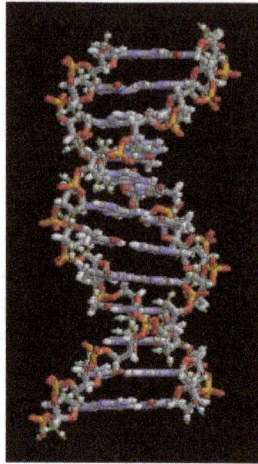

DNA structure. Bases are in the centre, surrounded
by phosphate–sugar chains in a double helix.

Recent findings have confirmed important examples of heritable changes that cannot be explained by direct agency of the DNA molecule. These phenomena are classed as epigenetic inheritance systems that are causally or independently evolving over genes. Research into modes and mechanisms of epigenetic inheritance is still in its scientific infancy, however, this area of research has attracted much recent activity as it broadens the scope of heritability and evolutionary biology in general. DNA methylation marking chromatin, self-sustaining metabolic loops, gene silencing by RNA interference, and the three dimensional conformation of proteins (such as prions) are areas where epigenetic inheritance systems have been discovered at the organismic level. Heritability may also occur at even larger scales. For example, ecological inheritance through the process of niche construction is defined by the regular and repeated activities of organisms in their environment. This generates a legacy of effect that modifies and feeds back into the selection regime of subsequent generations. Descendants inherit genes plus environmental characteristics generated by the ecological actions of ancestors. Other examples of heritability in evolution that are not under the direct control of genes include the inheritance of cultural traits, group heritability, and symbiogenesis. These examples of heritability that operate above the gene are covered broadly under the title of multilevel or hierarchical selection, which has been a subject of intense debate in the history of evolutionary science.

Relation to Theory of Evolution

When Charles Darwin proposed his theory of evolution in 1859, one of its major problems was the lack of an underlying mechanism for heredity. Darwin believed in a mix of blending inheritance and the inheritance of acquired traits (pangenesis). Blending inheritance would lead to uniformity across populations in only a few generations and then would remove variation from a population on which natural selection could act. This led to Darwin adopting some Lamarckian ideas in later editions of On the Origin of Species and his later biological works. Darwin's primary approach to heredity was to outline how it appeared to work (noticing that traits that were not expressed explicitly in the parent at the time of reproduction could be inherited, that certain traits could be sex-linked, etc.) rather than suggesting mechanisms.

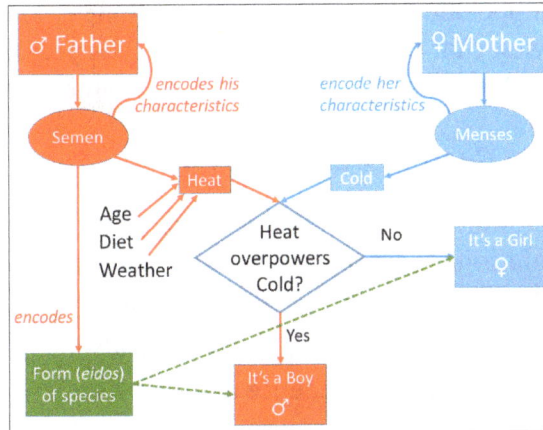

Aristotle's model of inheritance. The heat/cold part
is largely symmetrical, though influenced on the
father's side by other factors; but the form part is not.

Darwin's initial model of heredity was adopted by, and then heavily modified by, his cousin Francis Galton, who laid the framework for the biometric school of heredity. Galton found no evidence to support the aspects of Darwin's pangenesis model, which relied on acquired traits.

The inheritance of acquired traits was shown to have little basis in the 1880s when August Weismann cut the tails off many generations of mice and found that their offspring continued to develop tails.

Gregor Mendel: Father of Genetics

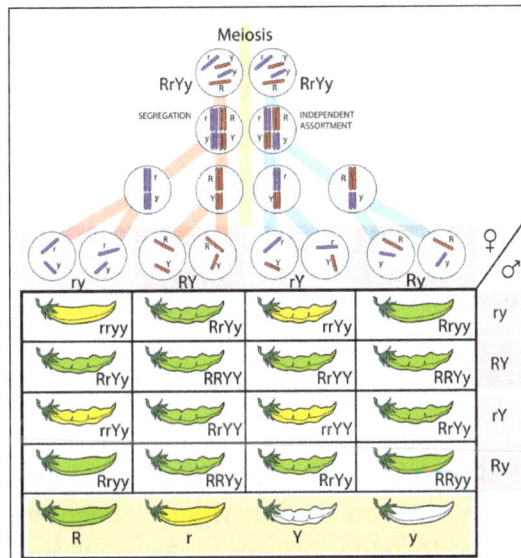

Table showing how the genes exchange according to segregation or independent
assortment during meiosis and how this translates into Mendel's laws.

The idea of particulate inheritance of genes can be attributed to the Moravian monk Gregor Mendel who published his work on pea plants in 1865. However, his work was not widely known and was rediscovered in 1901. It was initially assumed that Mendelian inheritance only accounted for large (qualitative) differences, such as those seen by Mendel in his pea plants – and the idea of additive effect of (quantitative) genes was not realised until R.A. Fisher's (1918) paper, "The Correlation

Between Relatives on the Supposition of Mendelian Inheritance" Mendel's overall contribution gave scientists a useful overview that traits were inheritable. His pea plant demonstration became the foundation of the study of Mendelian Traits. These traits can be traced on a single locus.

Modern Development of Genetics and Heredity

In the 1930s, work by Fisher and others resulted in a combination of Mendelian and biometric schools into the modern evolutionary synthesis. The modern synthesis bridged the gap between experimental geneticists and naturalists; and between both and palaeontologists, stating that:

1. All evolutionary phenomena can be explained in a way consistent with known genetic mechanisms and the observational evidence of naturalists.

2. Evolution is gradual: small genetic changes, recombination ordered by natural selection. Discontinuities amongst species (or other taxa) are explained as originating gradually through geographical separation and extinction (not saltation).

3. Selection is overwhelmingly the main mechanism of change; even slight advantages are important when continued. The object of selection is the phenotype in its surrounding environment. The role of genetic drift is equivocal; though strongly supported initially by Dobzhansky, it was downgraded later as results from ecological genetics were obtained.

4. The primacy of population thinking: the genetic diversity carried in natural populations is a key factor in evolution. The strength of natural selection in the wild was greater than expected; the effect of ecological factors such as niche occupation and the significance of barriers to gene flow are all important.

The idea that speciation occurs after populations are reproductively isolated has been much debated. In plants, polyploidy must be included in any view of speciation. Formulations such as 'evolution consists primarily of changes in the frequencies of alleles between one generation and another' were proposed rather later. The traditional view is that developmental biology ('evo-devo') played little part in the synthesis, but an account of Gavin de Beer's work by Stephen Jay Gould suggests he may be an exception.

Almost all aspects of the synthesis have been challenged at times, with varying degrees of success. There is no doubt, however, that the synthesis was a great landmark in evolutionary biology. It cleared up many confusions, and was directly responsible for stimulating a great deal of research in the post-World War II era.

Trofim Lysenko however caused a backlash of what is now called Lysenkoism in the Soviet Union when he emphasised Lamarckian ideas on the inheritance of acquired traits. This movement affected agricultural research and led to food shortages in the 1960s and seriously affected the USSR.

There is growing evidence that there is transgenerational inheritance of epigenetic changes in humans and other animals.

Common Genetic Disorders

1. Down syndrome
2. Sickle cell disease

3. Phenylketonuria (PKU)

4. Haemophilia

Types

Dominant and Recessive Alleles

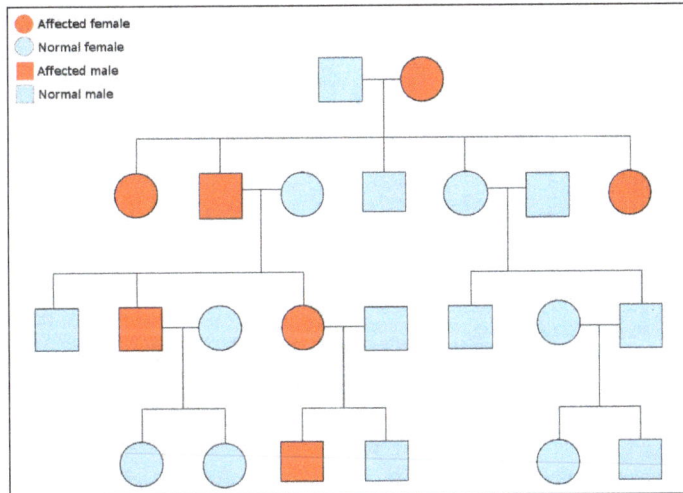

An example pedigree chart of an autosomal dominant disorder.

An allele is said to be dominant if it is always expressed in the appearance of an organism (phenotype) provided that at least one copy of it is present. For example, in peas the allele for green pods, G, is dominant to that for yellow pods, g. Thus pea plants with the pair of alleles either GG (homozygote) or Gg (heterozygote) will have green pods. The allele for yellow pods is recessive. The effects of this allele are only seen when it is present in both chromosomes, gg (homozygote).

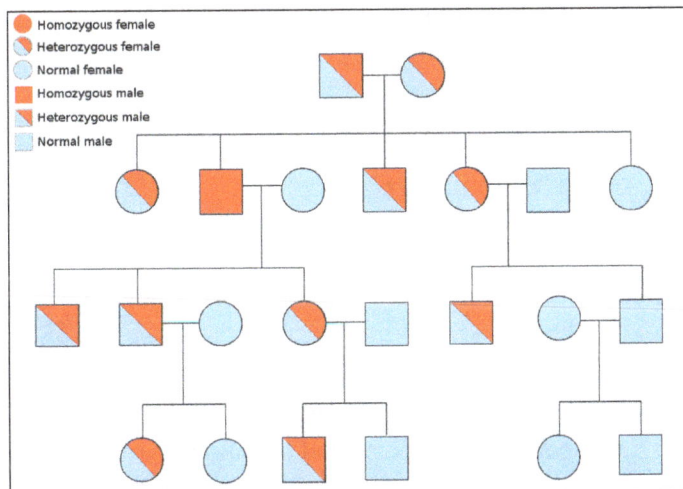

An example pedigree chart of an autosomal recessive disorder.

The description of a mode of biological inheritance consists of three main categories:

1. Number of involved loci

• Monogenetic (also called "simple") – one locus

- Oligogenetic – few loci

- Polygenetic – many loci

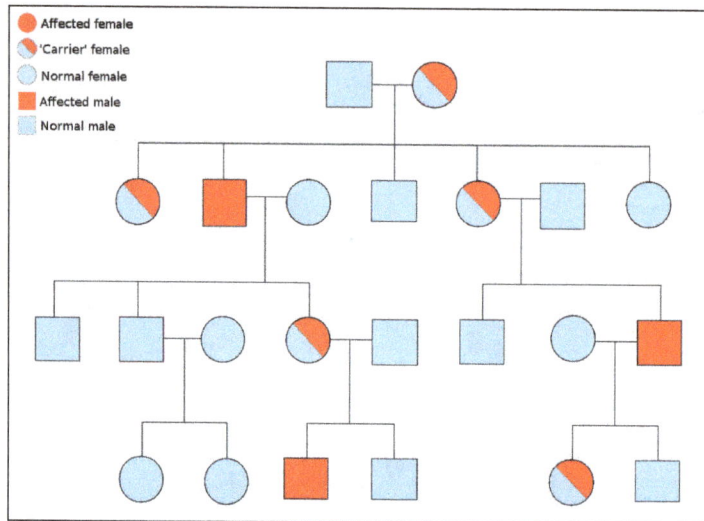

An example pedigree chart of a sex-linked
disorder (the gene is on the X chromosome).

2. Involved chromosomes

- Autosomal – loci are not situated on a sex chromosome

- Gonosomal – loci are situated on a sex chromosome

  ◦ X-chromosomal – loci are situated on the X-chromosome (the more common case)

  ◦ Y-chromosomal – loci are situated on the Y-chromosome

- Mitochondrial – loci are situated on the mitochondrial DNA

3. Correlation genotype–phenotype

- Dominant

- Intermediate (also called "codominant")

- Recessive

- Overdominant

- Underdominant

These three categories are part of every exact description of a mode of inheritance in the above order. In addition, more specifications may be added as follows:

4. Coincidental and environmental interactions

- Penetrance

  ◦ Complete

  ◦ Incomplete (percentual number)

- Expressivity
  - Invariable
  - Variable
  - Heritability (in polygenetic and sometimes also in oligogenetic modes of inheritance)
  - Maternal or paternal imprinting phenomena

5. Sex-linked interactions

- Sex-linked inheritance (gonosomal loci)
- Sex-limited phenotype expression (e.g., cryptorchism)
- Inheritance through the maternal line (in case of mitochondrial DNA loci)
- Inheritance through the paternal line (in case of Y-chromosomal loci)

6. Locus–locus interactions

- Epistasis with other loci (e.g., overdominance)
- Gene coupling with other loci
- Homozygotous lethal factors
- Semi-lethal factors

Determination and description of a mode of inheritance is also achieved primarily through statistical analysis of pedigree data. In case the involved loci are known, methods of molecular genetics can also be employed.

Genotype

Here the relation between genotype and phenotype is illustrated, using a Punnett square, for the character of petal colour in a pea plant. The letters B and b represent alleles for colour and the pictures show the resultant flowers.

The genotype is the part of the genetic makeup of a cell, and therefore of any individual, which determines one of its characteristics (phenotype). The term was coined by the Danish botanist, plant physiologist and geneticist Wilhelm Johannsen in 1903.

Genotype is one of three factors that determine phenotype, along with inherited factors epigenetic factors and non-inherited environmental factors. Not all organisms with the same genotype look or act the same way because appearance and behavior are modified by environmental and growing conditions. Likewise, not all organisms that look alike necessarily have the same genotype.

One's genotype differs subtly from one's genomic sequence, because it refers to how an individual differs or is specialized within a group of individuals or a species. So, typically, one refers to an individual's genotype with regard to a particular gene of interest and the combination of alleles the individual carries. Genotypes are often denoted with letters, for example Bb, where B stands for one allele and b for another.

Somatic mutations which are acquired rather than inherited, such as those in cancers, are not part of the individual's genotype. Hence, scientists and physicians sometimes talk about the genotype of a particular cancer, that is, of the disease as distinct from the diseased.

An example of a characteristic determined by a genotype is the petal color in a pea plant. The collection of all genetic possibilities for a single trait are called alleles; two alleles for petal color are purple and white.

Phenotype

Any given gene will usually cause an observable change in an organism, known as the phenotype. The terms genotype and phenotype are distinct for at least two reasons:

- To distinguish the source of an observer's knowledge (one can know about genotype by observing DNA; one can know about phenotype by observing outward appearance of an organism).

- Genotype and phenotype are not always directly correlated. Some genes only express a given phenotype in certain environmental conditions. Conversely, some phenotypes could be the result of multiple genotypes. The genotype is commonly mixed up with the phenotype which describes the end result of both the genetic and the environmental factors giving the observed expression (e.g. blue eyes, hair color, or various hereditary diseases).

A simple example to illustrate genotype as distinct from phenotype is the flower colour in pea plants . There are three available genotypes, PP (homozygous dominant), Pp (heterozygous), and pp (homozygous recessive). All three have different genotypes but the first two have the same phenotype (purple) as distinct from the third (white).

A more technical example to illustrate genotype is the single-nucleotide polymorphism or SNP. A SNP occurs when corresponding sequences of DNA from different individuals differ at one DNA base, for example where the sequence AAGCCTA changes to AAGCTTA. This contains two alleles: C and T. SNPs typically have three genotypes, denoted generically AA Aa and aa. In the example above, the three genotypes would be CC, CT and TT. Other types of genetic marker, such as micro-satellites, can have more than two alleles, and thus many different genotypes.

Penetrance is the proportion of individuals showing a specified genotype in their phenotype under a given set of environmental conditions.

Mendelian Inheritance

The distinction between genotype and phenotype is commonly experienced when studying family patterns for certain hereditary diseases or conditions, for example, hemophilia. Humans and most animals are diploid; thus there are two alleles for any given gene. These alleles can be the same (homozygous) or different (heterozygous), depending on the individual. With a dominant allele, the offspring is guaranteed to inherit the trait in question irrespective of the second allele.

In the case of an albino with a recessive allele (aa), the phenotype depends upon the other allele (Aa, aA, aa or AA). An affected person mating with a heterozygous individual (Aa or aA, also carrier) there is a 50-50 chance the offspring will be albino's phenotype. If a heterozygote mates with another heterozygote, there is 75% chance passing the gene on and only a 25% chance that the gene will be displayed. A homozygous dominant (AA) individual has a normal phenotype and no risk of abnormal offspring. A homozygous recessive individual has an abnormal phenotype and is guaranteed to pass the abnormal gene onto offspring.

In the case of hemophilia, it is sex-linked thus only carried on the X chromosome. Only females can be a carrier in which the abnormality is not displayed. This woman has a normal phenotype, but runs a 50-50 chance, with an unaffected partner, of passing her abnormal gene on to her offspring. If she mated with a man with haemophilia (another carrier) there would be a 75% chance of passing on the gene.

Determination

Genotyping is the process of elucidating the genotype of an individual with a biological assay. Also known as a genotypic assay, techniques include PCR, DNA fragment analysis, allele specific oligonucleotide (ASO) probes, DNA sequencing, and nucleic acid hybridization to DNA microarrays or beads. Several common genotyping techniques include restriction fragment length polymorphism (RFLP), terminal restriction fragment length polymorphism (t-RFLP), amplified fragment length polymorphism (AFLP), and multiplex ligation-dependent probe amplification (MLPA).

DNA fragment analysis can also be used to determine such disease causing genetics aberrations as microsatellite instability (*MSI*), *trisomy* or aneuploidy, and loss of heterozygosity (*LOH*). MSI and LOH in particular have been associated with cancer cell genotypes for colon, breast and cervical cancer.

The most common chromosomal aneuploidy is a trisomy of chromosome 21 which manifests itself as Down syndrome. Current technological limitations typically allow only a fraction of an individual's genotype to be determined efficiently.

Genetic Variation

Genetic variation is the difference in DNA among individuals. There are multiple sources of genetic variation, including mutation and genetic recombination.

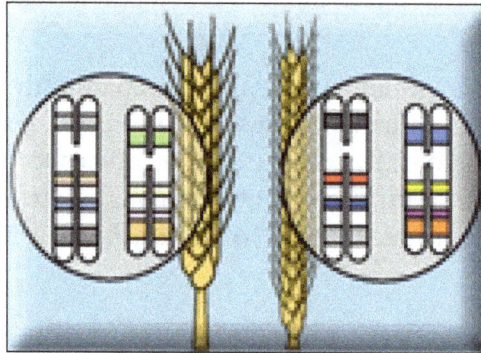

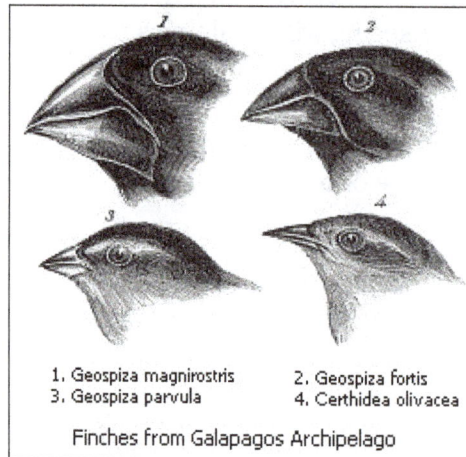

1. Geospiza magnirostris   2. Geospiza fortis
3. Geospiza parvula   4. Certhidea olivacea

Finches from Galapagos Archipelago

Darwin's finches or Galapagos finches.

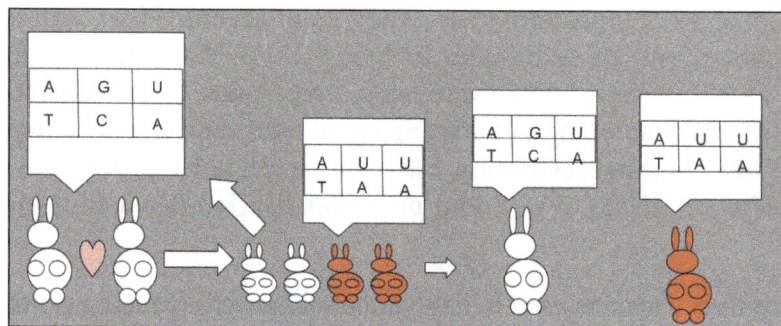

Parents have similar gene coding in this specific situation where they reproduce and variation in the offspring is seen. Offspring containing the variation also reproduce and passes down traits to their offspring.

Among Individuals within a Population

Genetic variation can be identified at a many levels. It is possible to identify genetic variation from observations of phenotypic variation in either quantitative traits (traits that vary continuously and are coded for by many genes (e.g., leg length in dogs)) or discrete traits (traits that fall into discrete categories and are coded for by one or a few genes (e.g., white, pink, red petal color in certain flowers)).

Genetic variation can also be identified by examining variation at the level of enzymes using the process of protein electrophoresis. Polymorphic genes have more than one allele at each locus. Half of the genes that code for enzymes in insects and plants may be polymorphic, whereas polymorphisms are less common among vertebrates.

Ultimately, genetic variation is caused by variation in the order of bases in the nucleotides in genes. New technology now allows scientists to directly sequence DNA which has identified even more genetic variation than was previously detected by protein electrophoresis. Examination of DNA has shown genetic variation in both coding regions and in the non-coding intron region of genes.

Genetic variation will result in phenotypic variation if variation in the order of nucleotides in the DNA sequence results in a difference in the order of amino acids in proteins coded by that DNA sequence, and if the resultant differences in amino acid sequence influence the shape, and thus the function of the enzyme.

Between Populations

Geographic variation means genetic differences in populations from different locations. This is caused by natural selection or genetic drift.

Measurement

Genetic variation within a population is commonly measured as the percentage of gene loci that are polymorphic or the percentage of gene loci in individuals that are heterozygous.

Sources

A range of variability in the mussel Donax variabilis.

Random mutations are the ultimate source of genetic variation. Mutations are likely to be rare and most mutations are neutral or deleterious, but in some instances, the new alleles can be favored by natural selection.

Polyploidy is an example of chromosomal mutation. Polyploidy is a condition wherein organisms have three or more sets of genetic variation (3n or more).

Crossing over (genetic recombination) and random segregation during meiosis can result in the production of new alleles or new combinations of alleles. Furthermore, random fertilization also contributes to variation.

Variation and recombination can be facilitated by transposable genetic elements, endogenous retroviruses, LINEs, SINEs, etc.

For a given genome of a multicellular organism, genetic variation may be acquired in somatic cells or inherited through the germline.

Forms

Genetic variation can be divided into different forms according to the size and type of genomic variation underpinning genetic change. Small-scale sequence variation (<1 kilobase, kb) includes base-pair substitution and indels. Large-scale structural variation (>1 kb) can be either copy number variation (loss or gain), or chromosomal rearrangement (translocation, inversion, or Segmental acquired uniparental disomy). Genetic variation and recombination by transposable elements and endogenous retroviruses sometimes is supplemented by a variety of persistent viruses and their defectives which generate genetic novelty in host genomes. Numerical variation in whole chromosomes or genomes can be either polyploidy or aneuploidy.

Maintenance in Populations

A variety of factors maintain genetic variation in populations. Potentially harmful recessive alleles can be hidden from selection in the heterozygous individuals in populations of diploid organisms (recessive alleles are only expressed in the less common homozygous individuals). Natural selection can also maintain genetic variation in balanced polymorphisms. Balanced polymorphisms may occur when heterozygotes are favored or when selection is frequency dependent.

Modern Evolutionary Synthesis

The Modern Synthesis describes the fusion (merger) of Mendelian genetics with Darwinian evolution that resulted in a unified theory of evolution. It is sometimes referred to as the Neo-Darwinian theory. The Modern Synthesis was developed by a number of now-legendary evolutionary biologists in the 1930s and 1940s.

The Modern Synthesis introduced several changes in how evolution and evolutionary processes were conceived. It proposed a new definition of evolution as "changes in allele frequencies within populations," thus emphasizing the genetic basis of evolution. (Alleles are alternate forms of the same gene, characterized by differences in DNA sequence that result in the construction of proteins that differ in amino acid composition.) Four forces of evolution were identified as contributing to changes in allele frequencies. These are random genetic drift, gene flow, mutation pressure, and natural selection. Of these, natural selection—by which the best-adapted organisms have the highest survival rate—is the only evolutionary force that makes organisms better adapted to their environments. Genetic drift describes random changes in allele frequencies in a population. It is particularly powerful in small populations. Gene flow describes allele frequency changes due to the immigration and emigration of individuals from a population. Mutation is a weak evolutionary force but is crucial because all genetic variation arises originally from mutation, alterations in the DNA sequences resulting from errors during replication or other factors. The Modern Synthesis recognized that the majority of mutations are deleterious (have a harmful effect), and that mutations that are advantageous usually have a small phenotypic effect. Advantageous mutations may be incorporated into the population through the process of natural selection. Changes in species therefore occur gradually through the accumulation of small changes. The large differences that are observed between species involve gradual change over extensive time periods. Speciation (the formation of new species) results from the evolution of reproductive isolation, often during a period of allopatry, in which two populations are isolated from one another.

There are several differences between the Modern Synthesis and the older Darwinian conception of evolution. First, mechanisms of evolution other than natural selection are recognized as playing important roles. Second, the Modern Synthesis succeeds in explaining the persistence of genetic variation, a problem that Charles Darwin struggled with. The dominant genetic theory of Darwin's time was blending inheritance, in which offspring were thought to be the genetic intermediates (in-between versions) of their two parents. As Darwin correctly recognized, blending inheritance would result in the rapid end of genetic variation within a population, giving natural selection no material to work with. Incorporating Gregor Mendel's particulate theory of inheritance, in which the alleles of a gene remain separate instead of merging, solves this problem.

There were several key players involved in the Modern Synthesis. The theory relied on the population genetics work of R. A. Fisher and Sewall Wright. Theodosius Dobzhansky made extensive studies of natural populations of the fruitfly *Drosophila* that supported many aspects of the theory. Ernst Mayr developed the biological species concept and created models concerning how speciation occurs. George Gaylord Simpson helped integrate paleontological observations into the theory behind the Modern Synthesis. G. Ledyard Stebbins contributed tenets (principles) based on his botanical work.

Since the 1990s it has been recognized that the Modern Synthesis omits some biological disciplines that are also relevant to evolution. In particular, much attention has focused on patterns of ontogeny and development.

References

- Gavrilets, s. (2004), fitness landscapes and the origin of species, princeton university press, isbn 978-0-691-11983-0

- Henry, devin (september 2006). "aristotle on the mechanism of inheritance". Journal of the history of biology. 39 (3): 425–455. Doi:10.1007/s10739-005-3058-y

- Bowler, peter j. (2003). Evolution: the history of an idea (3rd ed.). Berkeley, ca: university of california press. Pp. 129–134. Isbn 978-0-520-23693-6. Oclc 43091892

- Theobald, douglas l. (13 may 2010). "a formal test of the theory of universal common ancestry". Nature. 465 (7295): 219–222. Bibcode:2010natur.465..219t. Doi:10.1038/nature09014. Pmid 20463738

- ridley, mark. "speciation - what is the role of reinforcement in speciation?". Retrieved 2015-09-07. Adapted from evolution (2004), 3rd edition (malden, ma: blackwell publishing), isbn 978-1-4051-0345-9

- Pearson h (2006). "genetics: what is a gene?". Nature. 441 (7092): 398–401. Bibcode:2006natur.441..398p. Doi:10.1038/441398a. Pmid 16724031

- Allaby, michael, ed. (2009). A dictionary of zoology (3rd ed.). Oxford: oxford university press. Isbn 9780199233410. Oclc 260204631

- Albers p. K. And mcvean g. (2018): dating genomic variants and shared ancestry in population-scale sequencing data. Biorxiv: 416610. Doi:10.1101/416610

PERMISSIONS

INDEX

www.ingramcontent.com/pod-product-compliance
Lightning Source LLC
Chambersburg PA
CBHW082042190326
41458CB00010B/3440